T0267385

ATARI

ACORN

the

COMPUTERS

THAT MADE

BRITAIN

THE HOME COMPUTER REVOLUTION
OF THE 1980s

sinclair

First published in 2021 by Raspberry Pi
Trading Ltd, Maurice Wilkes Building,
St. John's Innovation Park, Cowley Road,
Cambridge, CB4 0DS

Publishing Director
Russell Barnes

Editors
Phil King, Simon Brew

Sub Editor
Nicola King

Design
Critical Media

Illustrations
Sam Alder with Brian O Halloran

CEO
Eben Upton

ISBN 978-1-912047-90-1

Contents

The chapters are arranged in order of each computer's availability in the UK,
as reflected by each model's date of review in Personal Computer World magazine.

Introduction

The 1980s was, categorically, the best decade ever. Not just because it gave us Duran Duran and E.T., not even because of the Sony Walkman. It was because the 1980s saw the rise of the personal computer.

With the help of hefty government discounts, computers inveigled their way into Britain's primary and secondary schools - even if teachers didn't always know what to do with them. Millions more computers appeared in living rooms and bedrooms around the country. For once, Britain was ahead of the world, helping to create a golden generation of British programmers. Sure, far more of us were destined to spend hours playing *Elite* and *Chuckie Egg* rather than creating games of our own, but the combination of C64s, Spectrum 48Ks, and BBC Micros directly led to the creation of a booming British software industry that continues to this day.

The question that inspired this book, though, is how did these computers come to be? There was no cookie-cutter template to follow. Companies were genuinely making things up as they went along, often to brilliant effect. Every computer you read about here has a story that surprises, and it's almost always down to the people involved. That's why, as much as this book is a story about each computer's creation, it's also a story about the people that created them. Many of them British, many of them geniuses.

With billions of pounds up for grabs in this nascent industry, not everyone was on their best behaviour. Ego battled ego in a quest for fame and wealth, leading to betrayals, lost fortunes, and too many shattered dreams to count. Think J.R. and *Dallas*, but with silicon instead of oil. And the fens of England rather than the sunbaked plains of Texas.

This book tells the stories of 19 computers that each had an impact on Britain. I apologise if your favourite is missing – I would love to have covered Apricot's machines, the NewBrain, the Oric-1, the Jupiter Ace, and the Cambridge Z88, for instance, but those will have to wait their turn.

While I fully expect people to jump straight to the computers they owned, this book has been designed to be read in any order. You can choose to start your journey in the late 1970s with the Commodore PET and end with the Acorn Archimedes in 1987; you can trace the stunning rise of Amstrad, which ultimately led to its swallowing Sinclair's Spectrum business whole. It's entirely up to you.

My final note is on accuracy. These computers were built two generations ago, which has given rumours and half-truths plenty of time to gather respectability. Wherever possible, I have gone straight to the source: that means listening to oral histories, reading interviews, ploughing through historic documents, and referring to previously written books. But most of all I have, wherever possible, spoken to the people involved.

The result is as close to 'truth' as I can get, and if that means slaying myths that have no root in reality then all the better.

In fairness, these are stories that need no exaggeration to make them fascinating. I hope you enjoy reading them.

Tim Danton

Research
Machines 380Z

A niche in which
to survive

Too small to rival IBM; ambitions too big to remain a supplier of components. Those were the forces that drove two entrepreneurial Oxbridge graduates to create their first computer for schools.

To trace the story of the Research Machines 380Z, we need to travel back in time to 1973. This was the embryonic age of computing, when companies were selling electronics in kit form. (An era, incidentally, that was reborn by the Raspberry Pi 40 years later.) It was a time when ambitious British entrepreneurs could hold a conversation in the pub about starting an electronics business without being ridiculed.

As normal, though, there is no straight line between idea and finished product. Mike Fischer and Mike O'Regan started their first business together after building a brain wave analyser for rats. You did not read that incorrectly. 'I was working as a contract electrician for a company called Roussel Laboratories, a drug company,' says Fischer. 'They wanted a special piece of scientific equipment built, and they knew I could do that sort of thing. So I got the contract to build it. It was a rat's brain wave analyser.' He laughs at the memory. 'It was not a large market.'

With the rat money safely in the bank, the two Mikes decided to create a new company called Research Machines Limited. It was, on the face of it, a triumph of hope over reality: neither Mike had any business experience, with only £200 cash and a German typewriter between them (the pound sign achieved, O'Regan explains, by typing capital L, backspace, hyphen).

O'Regan remembers Fischer saying, 'There's this company IBM, which stands for International Business Machines. It's hugely successful. Maybe we can be the equivalent in the scientific market.' This was also the reason behind the slightly archaic name Research Machines.

However, much to Intel's annoyance, they traded using the name Sintel; O'Regan recalls a threatening letter from Intel's solicitors that accused Sintel of 'passing off' as Intel by instructing its staff to answer the phone, 'Good morning, it'S intel.' 'Needless to say, that was not the case,' says O'Regan, 'and our tiny business was not even in the same market as theirs. I can't remember if we replied or not, but anyway we didn't hear any more from the solicitors, though it was a bit scary for us startup innocents to receive such a letter.'

The two companies could hardly have been any different. Where Intel was already a world-renowned chip maker, having created the revolutionary 4004 microprocessor in 1971, Sintel started life as a mail-order business selling components via small ads in electronics magazines. Fischer's enthusiasm for tinkering meant it wasn't long before it was advertising project kits too; if you wanted to build your own digital clock back in the early 1970s, Sintel was the place to go.

There was only so much money to be made from selling kits, and the pair wanted to build 'something' more substantial. When Fischer read about the new Zilog Z80 microprocessor, he realised that thing was a computer. 'It was the 1st of April 1976,' says Fischer, 'which I remember because it was April Fool's Day. We decided that we would have a go at making a computer based on the Z80.'

They had enough money in the bank to buy parts for 250 computers and, in a bold move, that's exactly what they did. 'We realised that microcomputers were going to become mainstream in the business world. And I thought that there was an opportunity to get in there before the big people, but that opportunity wouldn't last,' says Fischer, adding that he was realistic enough to know that the likes of DEC, HP, and IBM would soon trample on them – although in reality it took IBM five years to catch up. 'Our plan was to go into that market for a few years and then go and do something else, something derivative.'

However, fate was to take a different turn when two advisors from the Berkshire County Council approached Sintel asking for advice on how to build a microcomputer for education. 'We freely gave them lots of advice, and then suggested "Why don't we build a computer and you see whether it's appropriate?".'

The two Mikes soon realised that education was a niche they could attack. 'We thought we'll survive a little bit longer if we specialise in one market and do some software for that market as well,' recalls Fischer. 'So for the first few years, we spent a lot of time at exhibitions explaining to people why we wouldn't sell them a computer.'

First, though, they had a system to build. Time to start prototyping. Where Apple had a garage, Research Machines had Fischer and his wife's bedroom. 'Our bed was a double mattress on the floor, so we pushed the bed up against the wall during the day and pulled out the wallpaper trestle table.' He then set to work assembling a working system using a wire wrap model: 'It's essentially a breadboard with billions of wires coming out of it. You could literally build what you wanted, debug it, and then get a circuit board printed.'

Having settled on the core hardware, there was the small matter of writing the firmware to ensure the computer actually worked. This was beyond even Fischer, who drafted in David Small: neurology researcher at the University of Oxford by day, gifted programmer by night. In fact, Fischer describes Small as 'the best programmer I've ever come across'.

'I was like Steve Jobs to his Wozniak,' adds Fischer. 'I gave David a vision of what I wanted the BIOS to do and how to do it, and David went off and wrote it.'

This was only possible because Small had access to a PDP-8 minicomputer. 'David found a piece of software on the PDP-8 that allows you to write machine code. But he couldn't test it; to do that, he had to bring it to us. He wrote 4kB of machine code and it had about five bugs in it,' recalls Fischer. 'He was just a brilliant programmer.' So brilliant, in fact, that the two Mikes gave Small a third share of the company (this later caused problems when they fell out over his role in the company, with Fischer and O'Regan eventually buying Small out).

Fischer's other University of Oxford links came in handy too, with a brilliant young Physics postgraduate called Bob Jarnot building an EPROM programming machine to burn the read-only memory onto the chips – a machine they couldn't afford to buy.

At the same time, their education momentum was building. Those two Berkshire Council advisors introduced them to other school IT advisors, who became Research Machines' first major customers. In particular, Fischer flags the importance of Bill Tagg in Hertfordshire, who had been championing the importance of computers in schools since the 1960s; indeed, Maths pupils of Tagg at Hatfield School had access to an Elliot 803 since 1963 [1].

'Our second big breakthrough was a guy called Derek Esterson from the Inner London Education Authority. And those were the leading IT people in schools in Britain, in those days. And so the fact they bought from us meant that lots of other local authorities felt it was safe to buy from us.'

With a working prototype built, Research Machines Limited was ready to find buyers. The summer of 1977 would have found either Mike behind a stand in an exhibition, hoping to find a school or university willing to take the gamble. And they did. 'We sold one to a lovely, trusting young man from Bournville College. He bought our first one, which we delivered on time in September 1977,' says Fischer.

The other thing Fischer could sell was a vision. 'I understood the key things when microprocessors arrived, which is that they were only ever going to get better because of Moore's law.' This law dates back to 1965, when Intel co-founder Gordon Moore predicted that the number of components (and transistors in particular) on integrated circuits would continue to double each year for the next decade.

Fischer seized on this idea to help guide his development process. 'I understood that when you design something that you wanted to have a several-year product life, and particularly a computer, you should take into account the fact that DRAMs would get cheaper and bigger, and everything would get cheaper, so expandability was a key issue.'

This foresight is one of the main reasons why the 380Z had such longevity: Fischer chose expensive components from the start, on the basis that they were costly now but would be 'exactly the right thing two years later. So although we shipped the first one in 1977, even in 1979 we had a machine which was more or less optimal for the market. We had a graphics card, we had disk drives, we had a large amount of memory, and we had 80 characters.'

Reviewers were also fans of the Research Machines 380Z, with Mike Dennis describing the boards as a 'work of art' in his review of the computer in the very first issue of legendary computer magazine Personal Computer World [2]. While not without criticisms – he describes the 'general standard of construction' as 'more adequate than elegant' – his upbeat verdict ended by saying the 'monitor ROM and software backup are *excellent*'.

But it was expensive: if you chose the fully loaded 380Z with 32kB of memory and a floppy disk drive, then you would pay £1,787 (plus 8% VAT). A more basic system with 16kB of RAM and the keyboard still cost £965, which translates to around £6,000 in 2020. No wonder that Dennis suggested those on a budget should buy the 280Z for £400: this version simply consisted of the CPU board and VDU board, 'fully built and tested'. However, it only included 4kB of RAM and a 1kB ROM.

By late 1977, the two Mikes had honed their sales pitch, as is clear from adverts of the time. Pitched as 'the tool for research and education', and with universities firmly in its sights, the advert promised: 'Having your own 380Z means an end to fighting the central operating system, immediate feedback of program bugs, no queueing, and a virtually unlimited computing budget.' [3]

The two Mikes also wanted to lure schools and colleges, promising that a 380Z 'is ideal for teaching BASIC and Cesil [the Computer Education in Schools Language created by ICL]. For A Level machine language instruction, the 380Z has the best software panel of any computer.' A bold claim, but the company didn't stop there. 'This enables a teacher to single-step through programs and observe the effects on registers and memory, using a single keystroke.' The 'integral VDU' was another big selling point, along with the fact that it could display graphics alongside both upper- and lower-case characters.

The adverts also worked hard to convince readers that the 380Z would deliver a real return on investment: 'Microcomputers are extremely good value,' they began. 'The outright purchase price of a 380Z installation… is about the same as the annual maintenance cost of a typical laboratory minicomputer. It is worth thinking about!'

While the ads were shouting about immediate benefits, Fischer had a long-term vision for the computer in education. 'People had the magic view that microcomputers would be magic for education, but I could do the math. I could see that there was no way, for a long, long time, that there will be enough computer access and tools to use them in the curriculum.' Instead, Fischer wanted to give children access to the tools that they would one day use in their jobs.

In subsequent products from RM (as the company was later known), that would lead to Fischer negotiating a deal directly with Microsoft's Scott Oki, then in charge of the company's international operations, to bundle MS-DOS and Microsoft's early word processor and spreadsheet applications with RM's computers for 'nearly nothing'. In return, Fischer's argument went, Microsoft would have a strategic advantage: as each new generation of workers entered the workplace, it was Microsoft's software they would want to use. It's an argument that continues to hold force even now, although Google is doing its best to disrupt it with Google Docs.

Even at the launch of the 380Z, Fischer was determined to think about the whole package rather than just providing hardware. 'We went to great lengths to give what was a fairly unique thing at the time called the software front panel, which allowed you to single-step in a beautiful way through machine code and really understand how the computer was going.'

Another early advantage for RM is that it produced a cheaper, network workstation version of the 380Z, called the 480Z. 'In schools, there's no way that they could afford ten computers, each with disk drives,' says Fischer. 'So we had designed an Ethernet-like network, but using the Zilog communications chip. It was like a one megabit per second Ethernet. And we got that into schools.' As a result, a secondary school could afford a 380Z with disk drives, and then share those resources around a class via a series of 480Z workstations.

Despite the fact the first 380Z shipped in 1977, it was still a regular sight in British schools throughout the 1980s. Indeed, only RM and Acorn were officially sanctioned for use in schools for many years – which brings us to the thorny topic of the BBC Micro. 'We were approached by the BBC to bid for their concept of a BBC computer,' reveals Fischer, 'but we felt the timescales and price they had in mind weren't achievable.'

As is well documented, that contract was eventually won by Acorn for its BBC Micro – see its full story, starting on page 92 – and that caused big problems for RM. 'It could have bankrupted us,' says Fischer, so it's little wonder that he resented competing with the BBC name at the time. However, he was full of admiration for what Acorn achieved. 'The guys at Acorn did some pretty neat work with what was a very early gate array chip, to keep the chip count down. They deserved the success they had for a while.'

Although the majority of 380Z sales went to education, O'Regan points out that significant sales also went to commercial customers. 'The 50th 380Z was sold to IBM itself,' he says, 'with a front-page article in Computer Weekly coyly referring to the customer as "a leading multinational mainframe manufacturer".' Another early adopter was British Aerospace and, later, sales of over a hundred each went to the GPO (the Post Office as was) and to the Department of Education itself.

And it's in the education sector where the Research Machines name lives on. Following the 380Z range, RM produced a series of personal computers in 1985, badged the RM Nimbus, with Microsoft Windows and networking optimisation being key differentiators from Acorn's offering. For many years, RM maintained a market share of over 30% in schools (both primary and secondary).

Indeed, the company that the two Mikes founded back in 1973 continues to flourish in the education sector and is the only British computer 'manufacturer' that still exists – even if it stopped making hardware several years ago, instead focusing on the software and services side of the market.

If you were to count the number of British schoolchildren that had touched Research Machines hardware, and continue to use its services, over the course of the past four decades, it would be well into the tens of millions. That, as a legacy, is hard to argue with.

What came next

The RM 380Z is unusual in that it doesn't have a list of direct descendants. The RM 280Z was actually launched at the same time, while the RM Link 480Z was designed to be an accompanying system.

RM 280Z

Release 1977   **Price** £400

In the late 1970s computers tended to be self-assembled, which is why Research Machines felt comfortable to release a version that consisted of the CPU board and video board on their own. Buyers would then assemble the rest of the components, from RAM to keyboard.

RM 480Z

Release 1982   **Price** from around £480

While the RM 480Z could be used on its own – it included a 4MHz Z80A processor and up to 256kB of memory – its full name of Link 480Z gives away RM's intentions. Thanks to its networking technology, a bunch of cheaper RM 480Z computers could access the disk space on a linked 380Z and effectively turn it into a file server. Perfect for classrooms and shared work.

Sources
Interviews with Mike Fischer and Mike O'Regan.

1. Peter Excell, *A Pioneer Initiative in School Computing, Resurrection*: The Bulletin of the Computer Conservation Society, Issue 6, Summer 1993
   cs.man.ac.uk/CCS/res/res06.htm#h
2. Mike Dennis, Research Machines 380Z, Personal Computer World, Volume 1 Issue 1, 1978, page 47
3. Research Machines advert, Personal Computer World, June 1979, page 39

Commodore
PET 2001

The computer that
changed the world

You're on a TV quiz show. With a million pounds at stake, you have ten seconds to name the 'father of personal computing'. Who do you reach for? Steve Wozniak? Steve Jobs? Bill Gates? Jack Tramiel? Whoever you decide upon, chances are that you don't say Chuck Peddle. By the end of the story behind the Commodore PET, you may well change your mind.

One thing is impossible to contest: Peddle was a genius. But like many geniuses, he often found it difficult to persuade those in authority to see things his way. Nor was he the most diplomatic of men, on one occasion ripping off the arm of a chair in his determination to make a demo happen as he intended. If he had written a CV detailing his 1970s career, it would be full of high-ranking positions at three of the biggest tech companies – Apple, Commodore, Motorola – but marked by short or interrupted tenures at each.

Fortunately for the development of computing, that combination of brain power, passion, and visionary thinking led directly to the creation of the Commodore PET. And with it, the first mainstream microcomputer.

Had history followed a smoother course, that computer would have been built by Motorola. Peddle joined the company in 1973, hired to help develop the 6800 microprocessor. In particular, together with engineer Bill Mensch he created the crucial input/output interface that turned the 6800 into something genuinely useful. Peddle then became the 6800's key salesman, demoing and selling it to Hewlett-Packard, Ford Motor Company, and Remington among many others.

He kept on hitting a hurdle during those demos: value for money. While the companies loved the 6800 and the capabilities it provided, they weren't so fond of its $300 price. Peddle listened to the complaints and started to work on a low-cost version of the 6800, but this move was not greeted with wild enthusiasm by his employers: 'I got a formal letter [from Motorola] saying you have to stop work on your low-cost microprocessor,' Peddle told the Computer History Museum in 2014 [1].

Rather than meekly turn the other cheek, Peddle responded with fire. 'I wrote a letter back to Motorola and said, that's called project abandonment. So all of the work I've done up until now belongs to me, and I will not do any more development work for you... I'm going to go do it for myself.'

Peddle stayed on at Motorola, spending his time teaching companies how to use the 6800 while simultaneously hunting out funding for his pet project. One chance meeting later and he found himself visiting a small semiconductor manufacturer based in rural Pennsylvania: MOS Technology. At that point, MOS made its living by designing and fabricating calculator chips, but its president John Paivinen needed little persuasion that microprocessors were the future. He invited Peddle to set up his own team within MOS.

Peddle resigned from Motorola, and brought some of the key engineers from the 6800 team with him – including Bill Mensch. To say Motorola was displeased is an understatement, and it would soon launch a lawsuit against Peddle and his band of CPU refugees. Just to add a bit more spice, MOS called its two new processors the 6501 and 6502, with the 6501 being a drop-in replacement for the 6800.

It's important to note that Peddle claimed the processors took no DNA from the 6800. He instructed everyone not to take any paperwork with them when they left Motorola; this chip would be effectively designed from scratch (unfortunately, one engineer would disobey this instruction and give added weight to the Motorola lawyers' later arguments). If anything, Peddle said, the 6502 owes more to the processor architecture of the highly successful PDP-11 minicomputer.

While some people called the 6502 a RISC (reduced instruction set computer), Peddle always disputed this description. 'It wasn't. It was a reduced instruction set machine before that became a popular term at Stanford.' Instead, Peddle saw it as a 'universal solvent'. 'It's just enough, and it's simple enough, and it's cheap enough that you can use it for anything.' Crucially, it was also fast. While the 6502 lacked some of the Intel 8080's advanced features – for example, it didn't support 16-bit operations – it could complete tasks just as quickly due in part to pipelining (where the chip could accept new data even while it was processing existing data).

By June 1975 the 6502's design was complete, manufacturing obstacles overcome, and MOS Technology unveiled the new microprocessor to the world at Wescon 75. Short for the Western Electronic Show and Convention, this was a huge deal in the nascent computing industry, and MOS Technology's 6502 the stand-out product. With the show organisers banning sales on the show floor, attendees interested in buying the new chip – on sale for a paradigm-shifting $25 – needed to visit a suite in a nearby hotel. It was packed.

Among the many visitors to MOS Technology's hotel suite were two young men going by the names of Steve Wozniak and Steve Jobs, both of whom took home a processor and set of manuals. Peddle believed these were a big influence on Woz when designing the Apple I, although even Wozniak's genius wasn't enough to solve all the teething problems – Peddle later visited the famous garage where Apple was then based to troubleshoot some design issues.

While the 6502 proved a big hit at Wescon 75, the surprise package in terms of MOS Technology's finances was the KIM-1. This single-board computer included a calculator-style keyboard and six seven-segment LEDs for output. While it looks more like a calculator than a computer, its low price of $245 proved hugely attractive to hobbyists and development teams. MOS Technology reputedly sold over seven thousand units, adding a useful revenue line to the company.

The money was especially useful because MOS Technology had two fights on its hands: the lawsuit from Motorola and dropping calculator revenue due to arch-rival Texas Instruments slashing its chip prices. In early 1976, the company's financial backer – Allen-Bradley – cut its ties, and effectively handed ownership back to the three original founders of MOS. While they knew they had a hit on their hands with the 6502, they now faced a big problem: without financial backing, they couldn't invest the huge amount of upfront cash required for a semiconductor company.

Jack Tramiel, Commodore's ferocious leader, sniffed an opportunity. He owed MOS Technology money for the calculator chips it had already supplied, and while that traditionally might sound like a problem, he saw it as an opportunity: it gave him leverage. What's more, buying a chipmaker fitted in perfectly with his strategy of vertical integration. One that had been forged, in part at least, by Texas Instruments squeezing his supply of chips as it entered the calculator market itself.

According to Tramiel, speaking at an event to mark the 25th anniversary of the Commodore 64 [2], he put in a call to MOS president John Paivinen. 'I called him, we met, they were in very bad financial shape, they were losing $120,000 a month and they needed help, and I decided to buy this company and turn it over to become strictly a Commodore supplier.'

What he didn't appreciate at this point was that MOS Technology was sitting on a potential goldmine. During that same event, Tramiel describes MOS as being more interested in solving engineering problems than making money, with the result that it had 'a hundred or two hundred jobs from different companies to develop products'.

To determine which products were worth investing in, he asked people to come into his office every half hour to explain what they were doing.

'Almost the last one to come in was Chuck Peddle, and he showed me a product called the KIM,' said Tramiel. 'The KIM was a board, a PC board; if you attach the keyboard to the television then it was actually a computer. And he told me his idea to integrate these three pieces into one box and it could be a computer.'

Rather than dismiss the idea, Tramiel asked his technology-savvy son Leonard – then a postgraduate at New York's Colombia University – to head to Pennsylvania and meet Peddle to suss out whether he was a man worth listening to. 'That was one of the stranger conversations I've ever had,' says Leonard Tramiel. 'I expected to have a technical discussion about microprocessors, and how he wanted to construct this thing and what the features would be.

'Instead we spoke for at least two-thirds of the discussion about a Robert Heinlein short story called *The Door Into Summer*, which was about a future society that was just littered with embedded microcomputers. That was the society that Chuck wanted to live in. And in order for people to be comfortable with such a society, he knew that personal computers would be necessary. And to do that, we would have to have very inexpensive, yet powerful microprocessors. So he had done the first step, which was to develop the 6502. And the next step was to make a personal computer.'

Leonard duly reported back to his father that Chuck Peddle wasn't crazy. 'He said how fantastic [the 6502] is,' said Jack Tramiel, 'so I gave Mr Peddle six months to come up with a prototype. If he does, he can stay on; if not, goodbye. And six months later, we went to a Chicago winter electronic fair, and we showed the product – it was an unbelievable success.'

That rather simplifies the development process. At one point, Jack Tramiel and Steve Jobs met to discuss the prospect of Commodore buying the young company and using its forthcoming Apple II system as the basis of the new computer's design, but the two men couldn't agree a price. Even if the deal had gone through, it's impossible to imagine Jobs working for such a strong personality as Jack Tramiel.

So Peddle and a small team of engineers set to work on their own prototype, guided by a specification from Radio Shack. The then-dominant American electronics retailer was keen to sell a fully built computer, and it wanted it to be so simple that franchisees could sell it without needing to offer support. They were looking for something 'turnkey', to borrow a phrase from Chuck Peddle. 'And so they published a spec that says we want a built-in CRT, we want a tape drive built in, we want people to be able to load the programs, and we want them to be able to run.'

Peddle's ambitions went far beyond what Radio Shack was thinking. For starters, he was determined to include BASIC so that people could sit down and start programming themselves; he knew there was a pent-up demand from people who were interested in coding, but less interested in building their own computers with welding irons and do-it-yourself kits.

By this point, Microsoft (or Micro-Soft as it was then known) was already the big name in BASIC, having convinced Altair to distribute and market their version of the language in the Altair 8800 microcomputer (see 'The computers that came before', page 35). That's why Peddle dropped by Microsoft's early offices in Albuquerque, New Mexico, and requested a version of BASIC for the Commodore.

In particular, Peddle wanted support for the IEEE 488 interface. Forty years later and it's hard to get excited by IEEE 488, but it allowed users to hook up peripherals such as printers, disk drives, tape drives, and scientific equipment. Until the PET, it was only found on HP minicomputers costing $5,000 and more, and it gave the computer what Peddle describes as 'dignity'.

'We're going to be selling this thing for $500, and everybody else thinks computers are $20,000 and it's got to have dignity,' Peddle told the Computer History Museum. 'So putting the IEEE 488 in gave me dignity, and also gave me a guaranteed market of people that would buy it… because I would be able to sell it against HP and Tektronix and everybody that were selling them for thousands of dollars.'

Bill Gates set to work on the BASIC without agreeing a price, which was always a mistake when dealing with Jack Tramiel. As the Microsoft founder discovered a few months later. '[Gates] came to see me and tried to sell me the BASIC, and he told me that I don't have to give him any money, I only need to give him $3 per unit,' said Tramiel in 2007. 'I told him that the highest price I'm willing to pay is $25,000 [as a lump sum], and about six weeks later he came and took the $25,000. And since then, he doesn't want to speak to me.'

Kit Spencer, then in charge of Commodore's marketing in the UK, remembers the night the deal was signed, and describes it as not only a great deal for Commodore – 'we got an operating system that we put on 20 million computers and it didn't cost us anything in reality' – but a crucial lesson for Bill Gates and Microsoft. With his fingers thoroughly burnt, Gates would later turn round to IBM and insist, in Spencer's words, 'you can have it for a tiny royalty and I'll keep it up to date for you. And he kept the rights to it. And that is the history of Microsoft.'

With the key software commissioned, Peddle and his team set to work on the hardware. Rather than start from scratch, they opted for a design not dissimilar to the KIM-1, already created for a computerised sprinkler control system. This isn't quite as bizarre as it seems. The system had been developed by Petr Sehnal, a colleague of Peddle's from MOS Technology, and Peddle remembers it as a 'little general-purpose thing that was designed for that kind of application' [3].

One big challenge was integrating the IEEE 488 slot, a responsibility that fell on the shoulders of Bill Seiler, one of Peddle's most trusted and gifted engineers. 'We figured out a way to implement it cheaply all in software as much as we could,' said Peddle in Brian Bagnall's *Commodore: A Company on the Edge*. 'The devices have to respond in a certain amount of time, so it created a little bit of a headache for us. We kind of blew it a little bit because it needed some hardware assist, but we didn't do that.'

To turn ideas into a working prototype, Seiler relied on Nobuo Aoji, a Japanese engineer who Tramiel had flown over to California from Japan. Seiler describes Aoji as 'like a Tasmanian devil in the lab' [4], able to translate one of his schematics into a working breadboard prototype within an hour.

Peddle asked Larry Hittle, who had made the KIM-1, to design the PET's case. Hittle went on to make two wooden prototype cases with smooth, rounded edges that hinted towards a futuristic machine. This wasn't a totally original concept, with similarly rounded Courier terminals already in existence – and including both a built-in CRT monitor and a keyboard.

While it was a stylish design, it never made the final cut as Tramiel couldn't afford to mass-produce the PET with such curves. Instead, and here's where Commodore's vertical integration made its presence felt once more, he instructed his filing cabinet factory in Toronto to use metal to create something that looked like the prototype.

This was an acceptable compromise, but Peddle conceded one other major design decision to Tramiel: the keyboard. 'We're showing him a typewriter keyboard because we say all computers need a typewriter keyboard, because we had all grown up with them like that,' Peddle told the Computer History Museum in 2014. Tramiel wasn't convinced and asked for evidence that this is what people wanted. Especially when Commodore was already making calculator keyboards. Peddle conceded the point, and the decision was made.

The final key component was a monitor, and once again the combined pressure of a tight deadline and an equally tight focus on costs meant a small, cheap model was the only option. Peddle despatched a young employee to a local electronics store to find such a screen, and he returned with a cheap, black and white TV. It was then a matter of taking it apart and making it work. With none of his team having experience with CRTs, they used a how-to guide to try to connect everything as it should. Which they did, except for one thing: when they switched it on, the screen image was upside down.

It was this final technical challenge that kept Peddle awake at a freezing CES Chicago in January 1977. With temperatures dropping below -20°C, and an ice fog across the lake, Peddle spent a final night 'trying to get things running to meet with Radio Shack'. Their contact was John Roach, who would go on to usher Radio Shack into the computer age.

'John Roach comes up, and we show him this machine, sort of working. I mean, I think we got the CRT right side up just in time,' said Peddle. 'And he and Tramiel had a discussion. Tramiel says I'll finish this for you, but in order to do that, you're going to carry my full calculator line.' Roach, who could see the fate of calculators better than Tramiel at that time, said no. 'So Tramiel said OK, well then you can't have my computer.'

While Roach's team went on to design the popular Tandy Radio Shack TRS-80 (the 80 refers to the fact it was based on a Zilog Z80 processor), Tramiel was sufficiently convinced that Commodore should invest in Peddle's microcomputer. It helped that, following a leak of the computer's design to Electronic Engineering Times, the company's stock rose from $4.50 to $7. From not being sure whether Peddle had a place in his company, Tramiel was now keen to motivate his chief engineer to even greater heights – promising him $1 for every computer sold.

Now the pace quickened. In March 1977, Commodore unveiled the world's first personal computer at the Hanover Fair. It would include 4kB of programmable memory, BASIC, a keyboard, a monitor, and a cassette drive for $495. By comparison,

Apple was soon advertising the Apple II – also based on the 6502 and with 4kB of RAM – for $1,298, and that was without a screen.

While the PET didn't generate many column inches at the time, there were already signs that it could be big. 'We had the first prototype on the stand next to Wang,' says Kit Spencer. 'On the first day, some of their sales guys came to look at it. Then the second day the general manager of Wang comes up. By about the fourth day of the show, the assistant to the president of Wang came to look at it.'

For Jack Tramiel, the buzz created by the prototype PET was enough to make him take action. 'A little while later, he said to me, I think this is the future,' says Spencer. 'This is the future of the company. He said, Kit, can you start a new division because I just want to find out what we can really do. We talked about the possibilities: could this replace typewriters, could it do this, that and the other. We knew there was incredible potential, but nobody knew what it could do.'

American buyers who wanted to see the PET for themselves had to wait until the West Coast Computer Faire in April 1977. This new show had been launched specifically with home computing in mind, with two-page adverts in Byte magazine that promised '100 Conference Presentations', '200 Commercial & Homebrew Exhibits', and 'Two Banquets with Outstanding Speakers' [5].

An estimated 13,000 people headed to the Civic Auditorium in downtown San Francisco, with Peddle describing the atmosphere as 'a lot like Woodstock' [6], with the usual suited computer conference attendee replaced by people in jeans. 'These were the guys with the thick glasses and the slide rules in their pocket. This was their thing.'

With the PET wrapped in its futuristic casing, and including everything a microcomputer user could wish for in one package, it stood out from the collection of boards seen elsewhere on the show floor. However, a note of caution: judging from Byte's coverage of the show [7], where it didn't mention the PET at all, we would be rewriting history to suggest that it was a colossal hit.

It's telling that the PET didn't even draw much attention at the summer edition of CES; in many ways, Commodore was ahead of the market. Computer dealers were comfortable selling machines with four-figure and five-figure price tags, and there was no proof yet that the PET would be any more successful than the hobbyist board systems that had gone before. Commodore had just one significant customer: a man who sold used computers in the Midwest, and who was so convinced by the

PET that he handed Peddle a cheque for $25,000. Once Commodore had shipped him that many computers, he said, he would send another cheque.

This was a sign of things to come. At the National Computer Conference in Dallas, Peddle was finally able to demonstrate a collection of working PETs rather than cobbled-together prototypes, and the response was overwhelming. With Tramiel having commanded Peddle to accept pre-orders – Commodore desperately needed the money to kick-start production – Peddle and his wife spent the next three days accepting cheques from ordinary punters desperate to receive their PETs. This was despite orders being limited to 8kB machines at $795.

'I walked upstairs almost at the end of the show to see what was going on with the big computer guys,' said Peddle [8]. 'There wasn't anybody on the floors up there. Everybody had left the floors and were downstairs buying computers. It was the end of the big computer time... we had absolutely wiped them out.'

By autumn 1977, momentum had swung almost entirely in Commodore's favour. US magazine Personal Computing put the PET on the cover of its September/ October issue, asking 'Is it the first of a new generation?' In a ten-page article that featured an extended interview with Chuck Peddle [9], he made many claims that have stood the test of the time – and others that haven't. For instance, it took some time for his prediction of the computer's role in the average person's planning finances to come true: 'He has his computer to do his planning; he approves the plan in his home, and transfers a cassette to the bank, letting the bank automatically do the paying.'

Peddle also predicted the role of computers in education. 'Many of the schools will now be able to teach children at ages 7, 8, and 9 the fundamentals of programming and the fundamentals of using a computer... It will actually teach lessons, allowing children in the schools to proceed at their own rate.' His hope that the whole nation would become programmers, able to whip up personalised programs to fit a purpose, proved less visionary.

As demand rose, Commodore struggled to keep up. With chairman and purse holder Irving Gould unwilling to gamble money on a huge production facility, the company made do with a rented factory where labourers painstakingly assembled the machines at desks. The first PETs were shipped in October, but even towards the end of the year they were making around 30 machines per day, with an estimated total of 500 by the end of 1977.

Peddle told the story of one lady who begged him to give her a PET early. 'She says, you've got to do this. We have one computer at home. There's four of us using it. We drew to see the time that we get to use the computer, and I'm the housewife, but my time is from 2 o'clock in the morning until 6 o'clock.'

Those lucky enough to receive the first batch of PETs weren't really so lucky. For example, early customers complained about the keyboard because the letters wore off and it was difficult to type on. Meanwhile, the TRS-80 was gaining momentum and Apple was catching up. Not only did the Apple II offer colour, but Wozniak had created a disk drive. Peddle was dismissive of Apple's single drive approach, wanting to engineer a dual drive which he felt was more professional, as it would allow people to save work to the second disk. However, it would take him months to do so.

Then there were distribution issues. While Radio Shack had a network of stores, Commodore could only rely on its much more limited Mr Calculator outlets, while Tramiel's refusal to give ComputerLand a discount meant they decided not to stock any PETs.

Instead, he looked to Europe. After dodging one immediate legal hurdle – Philips had the rights to the name PET, so Commodore rebranded it as the CBM 3000 series – he saw the opportunity to focus on profit. With only the less flexible Tandy TRS-80 as a rival, he set a high price of £695. This was equivalent to $1,295, so effectively British customers were being asked to pay almost twice as much as Americans were paying. To look at it another way, Brits were being asked to pay £4,000 when adjusted for inflation (the PET arrived on UK shores in early 1978).

You might think this would irritate the man charged with building sales in the UK, especially as this encouraged a 'grey market' of PETs making their way from the USA through illicit channels, but Kit Spencer is philosophical. 'It was an irritant, but it didn't stop us getting the market going.'

It helped that Spencer had a clear vision. In particular, he saw the UK launch of a fully-fledged computer as a chance to do things differently: to create a long-term business, complete with a proper infrastructure and support for buyers. And to make the PET user-friendly. 'I remember when we got the first half-dozen PETs, this complex product, they came over with no manuals or anything. So the first guy I hired, I said go home, go anywhere, just write me a manual. He managed to create a pretty crude manual, I think it was probably about 40 Xerox pages, and we put that in the first PET.'

Tramiel wasn't impressed when he discovered this, saying that it was the US company's job to produce the manual. ' "Why are you doing the manual?" he says,' recalls Spencer. 'I said, "Because we haven't got one, Jack." Jack says OK. The next day, I get a phone call from the US. "Jack tells me you've got a manual, can you send it please?"'

This was typical of Spencer's relationship with Jack Tramiel. While other Commodorians often fell out of favour with the company's passionate leader, Spencer had the advantage of being several thousand miles away – and also had results to back up his decisions. 'Jack left us very much to get on with things, which was good, because you're able to make plans without being micromanaged and really see what it could do. And I think we just maybe did make the right decisions – we certainly became the dominant computer in the UK and, in fact, virtually all of Europe.'

This freedom allowed Spencer to effectively create a rule book on how to launch a computer. 'I realised we had a car, so we had to provide the driving lessons. In the UK, we were able to put in a lot of support operations fairly quickly. They weren't big. I had someone doing the training division, I had somebody doing software. But I made them all profit centres, because Jack was always pretty hard on costs.'

Spencer also started, albeit in humble form, the first magazine built around a specific computer. 'Even before we sold a Commodore computer we had a prototype, and I could see the interest in it, so I decided to start a Commodore newsletter. I think we charged something like £10.' In return, Commodore UK promised, it would send you a monthly newsletter to keep you up to date.

The British found useful allies in higher education too, with Strathclyde University buying a number of Commodore PETs, and then using them to write software that would prove popular with users. If anything, this pattern carries on strongly to this day, with Britain producing a huge amount of innovative software packages ever since.

While this book focuses on the hardware, it's crucial to understand how important software has always been – even in the late 1970s – when people are deciding which computer to buy. While Peddle may have had the notion of every computer user being a programmer, in reality that's a tiny subset of the population: most people simply like to exploit the work produced by others, whether that's accounting software, a word processor, or a game. One of the early drivers of PET sales was *Microchess*, which Commodore was demoing in shows as early as 1977.

In 1978, estimates put UK sales of the Commodore PET at 30,000. It helped that Apple wasn't yet established in the UK, and Commodore also didn't have to fend off competition from the TRS-80. Instead, Spencer saw his role as developing the market. 'Tandy and Sinclair came in eventually, but for the first year or two there was not a lot of competition per se. The competition was traditional computers and convincing people we were useful.'

Douglas Adams, author of *The Hitch-Hiker's Guide to the Galaxy*, was one of the unconvinced. 'I remember the first time I ever saw a personal computer,' he would write in 1999 [10]. 'It was at Lasky's, on the Tottenham Court Road, and it was called a Commodore PET. It was quite a large pyramid shape, with a screen at the top about the size of a chocolate bar. I prowled around it for a while, fascinated. But it was no good. I couldn't for the life of me see any way in which a computer could be of any use in the life or work of a writer.'

As the market evolved, so did the Commodore PET. In late 1978, Commodore released new versions with a typewriter-style keyboard, while jettisoning the built-in cassette drive (it sold a standalone unit for $95). It also released a trio of printers, built in partnership with Epson, that cost up to $795. What the PET still didn't have, and Apple did, was a disk drive.

This would ultimately lead to Chuck Peddle leaving the company. Tramiel was furious that Commodore had been outflanked by Apple releasing its drive first, at a price of $595, and while it was basic and slow, it worked. At that time, Commodore had nothing. Peddle had been working on a dual disk drive for several months but technical, staffing, and health problems had all combined to derail the project. Tramiel demoted Peddle and took him off the project. With Apple promising him riches if he switched sides, and Tramiel only punishment, Peddle took the difficult decision to leave.

Peddle's months at Apple weren't happy ones, but his replacement soon discovered just how hard it was to create a floppy drive. When the chance arose for Peddle to rejoin Commodore in February 1979, he leapt at it. He brought the project back on track, and finally unveiled the 2040 dual drive at the June 1979 CES. It cost $1,295 and was more like a second computer than a peripheral, containing two processors, its own operating system, and 4kB of memory. Crucially, though, it didn't steal any precious computing power from the main computer, whereas Apple's drive did.

The Apple II had one glaring advantage over PET, however, and that was colour. While Commodore had continued to improve the PET during 1979, including a large-screened version with an 80-column display that was targeted at businesses, the PET's mono approach was starting to look old-fashioned. What's more, Apple had stolen a march thanks to the release of VisiCalc, the first spreadsheet.

Commodore needed to do something. If Chuck Peddle had won the argument, that something would have been the ColorPET: an out-and-out business machine costing well over a thousand dollars. To find out what actually happened, turn to the story of the VIC-20 (page 62).

How the PET got its name

The Commodore PET may well not have happened at all if it wasn't for Andre Sousan. At the time when Commodore bought MOS Technology, Sousan was VP of engineering – and a former colleague of Chuck Peddle at Texas Instruments.

Where Peddle was brash, Frenchman Sousan was smooth and sophisticated. Peddle soon convinced him of the merits of a true microcomputer, and it was Sousan and Peddle in combination who persuaded Tramiel that an all-in-one unit, complete with monitor, was the right way to go.

With Tramiel signed up, they now just needed to think of a name. 'At that time there was a phenomenon called the Pet Rock,' explains Michael Tomczyk, who would play an important role in the release of the VIC-20. 'Somebody was selling rocks in a box and calling it the Pet Rock. It was a big fad.'

The word 'pet' appealed to Sousan due to its warm, fuzzy nature – exactly the characteristics they wanted people to associate with their new computer. Now they had an acronym, they just had to create words to match. While 'personal' and 'electronic' were easy, the letter T proved more problematic. 'Chuck said he went home, he scanned through the dictionary and the next day he came back to Jack and he said we're going to call it the personal electronic transactor. So that's what the PET stands for and that's how they got over a potential lawsuit from the Pet Rock people.'

The 2001 is less romantic. While *2001: A Space Odyssey* was almost ten years old by the time the PET launched, it still epitomised the idea of space age futurism. And what could sound more futuristic than the Commodore PET 2001?

The computers that came before

Note that we aren't suggesting for a moment that the PET was the first computer. Colossus famously helped shorten the Second World War due to its code-breaking prowess, and throughout the 1950s, 1960s, and 1970s mainframe computers played an increasingly important role in business-critical applications.

Mainframes needed rooms dedicated to their needs, but with the arrival of the DEC PDP-8 minicomputer in 1965 things started to change. While hardly cheap at $18,500 ($150,000 in today's money, according to the US Inflation Calculator), its size and relative affordability helped it sell almost 50,000 units. With the PDP-11, released in 1970, things got even more serious: according to its entry on Wikipedia, DEC sold over 600,000 PDP-11 minicomputers across its 20-year lifespan.

But the computer that predates the Commodore PET and has the strongest claim to be the first microcomputer is the Altair 8800. It used a 2MHz Intel 8080 processor and was initially only available in kit form: it was designed to appeal to hobbyists and electronics fans, not the general public. It was comparably cheap too, at $395.

It was the Altair that inspired many computer pioneers into their trade: most notably, Microsoft wrote a version of BASIC for the computer that put them in business, and it was the subject of the first meeting of the Homebrew Computer Club (which spurred Steve Wozniak to create the Apple I).

While it's inevitable that personal computers would have taken off without the Altair – and note that it barely sold at all in the UK – it was the spark that lit up a whole industry.

Beyond ASCII

One of the PET's most loved features was its extended character set. By pressing Shift followed by the 'A' key, you would bring up a spades icon from a deck of cards. And, naturally, programmers had access to all these special characters too, allowing them to generate graphically interesting games by using 'normal' characters.

Commodore fans christened this character set PETSCII, which is derived from ASCII – short for the American Standard Code for Information Interchange. PETSCII (not that he approves of the name) was created by Leonard Tramiel, son of Jack, in the months before the PET's release. His brief from Chuck Peddle? 'It must play cards. There's 256 symbols in an 8-bit character set. Basically 64 of them are fixed because of letters and numbers and punctuation and the like. He gave me the mandate that I must include hearts, spades, diamonds, and clubs, so that he could play blackjack. And everything else was just open. Figure out what will work.'

To help him decide what to include, Tramiel had two images in his mind: the Starship Enterprise and the Apollo Lunar Module. 'They were both in my mind full-screen. So they were much larger than the *Lunar Lander* game, but by the time I was done designing the set, I had something that made a pretty darn good lunar lander. So I thought, "Oh, I can write a game around this." And I did.'

Unlike Apple with the II, it was also possible to type in lower-case letters. Although, admittedly, not totally straightforward. First you had to invoke the relevant POKE command: 'POKE 59468,14'. Even then, to get a lower-case letter you need to press Shift first. When you were done, you flipped back into graphics mode by typing 'POKE 59468,12'.

To have a play yourself, head to **masswerk.at/pet**.

What came next

PET 2001-N, PET 2001-B

Release 1978 **Price** from £795

After receiving endless complaints about the calculator-style keyboard of the original PET 2001, Commodore released two models with a full-size keyboard – and dropped the integrated cassette drive. The 2001-N included graphical PETSCII characters, while the business-focused 2001-B just had regular characters.

PET 4000 series/CBM 8000 series
Release 1980   **Price** from £1,295

The 4000 series' key selling point was its larger screen, complete with support for 80-column text, but unlike the previous PET you couldn't upgrade the RAM yourself: you had to choose 8kB (the 4008), 16kB (4016), or 32kB (4032) and live with it. While the CPU stayed the same, it ran faster due to improved circuitry and included an extended version of Commodore BASIC.

SuperPET 9000 series/MicroMainframe
Release 1981   **Price** from £2,000

Designed for university students, programmers, and scientists, the SP9000 included 96kB of RAM and – thanks to an RS-232 interface – the ability to send work back to a mainframe (perfect for students). The 48kB ROM also included a bumper collection of languages, including COBOL, Fortran, and Pascal.

Sources
Interviews with Kit Spencer, Leonard Tramiel, and Michael Tomczyk.

1. Oral History of Charles Ingerham 'Chuck' Peddle, Computer History Museum, 12 June 2014, **youtu.be/enHF9lMseP8**
2. Impact of the Commodore 64: a 25th anniversary celebration, Computer History Museum, 7 December 2007 **computerhistory.org/collections/catalog/102695068**
3. Brian Bagnall, *Commodore: A Company On The Edge*, Kindle edition, location 1473
4. Brian Bagnall, *Commodore: A Company On The Edge*, Kindle edition, location 1814
5. Byte Magazine, Volume 2, Number 4, April 1977, page 68 **archive.org/details/byte-magazine-1977-04/page/n69**
6. Brian Bagnall, *Commodore: A Company On The Edge*, Kindle edition, location 1930
7. Byte Magazine, Volume 2, Number 7, page 25 **archive.org/details/byte-magazine-1977-07/page/n25**
8. Oral History of Charles Ingerham 'Chuck' Peddle, Computer History Museum, 12 June 2014, **youtu.be/enHF9lMseP8**
9. Tom Munnecke, 'Chuck Peddle On The PET Computer', Personal Computing, September/October 1977, page 30 **archive.org/details/PersonalComputing19770910/page/n29**
10. Douglas Adams, *The Salmon of Doubt*, Pan Books 2003 edition, page 91

Apple II

The birth of
a superpower

To understand the birth of the Apple II, often stylised the Apple][, you need to transport yourself back to California in 1975. Two young men: one an engineer, one a natural-born salesman. And both called Steve.

At that time, Steve Wozniak was a loyal employee of Hewlett-Packard with a deep love of tinkering with computers. Steve Jobs was a highly intelligent college dropout with a curious moral compass, despite his keen interest in spiritualism. For example, while working for Atari, Jobs told Wozniak that he would split the $700 fee for a board redesign challenge if Wozniak would do the work. Wozniak did such a phenomenal job that Atari paid a bonus of $5,000, but Jobs only gave his friend $350.

Wozniak, universally called Woz, had been tinkering with electronics since he was a child. 'A lot of things I did back then I'd probably be put in prison for a few years now,' he said in a lecture at San Francisco's Computer History Museum back in 2002 [1]. 'One time I built a little electronic metronome and put it in a friend's locker, and it was going "tick, tick, tick" and it had these big battery cells that kind of made it look like a bomb,' he said.

Unfortunately, it wasn't his friend that found it but the school principal. 'I had to laugh when the principal told me how he opened up the locker, clasped it to his chest, and ran out and dismantled it. But because I had rigged it with a resistor, when he opened the locker the ticking sped up.'

So possibly not quite as funny for the principal. More seriously, the young Woz spent much of his time at high school designing computers on paper, based on the chip technology available at the time. When a new chip appeared, he, for fun, would redesign his existing computers to take less space or use fewer chips.

There could be no better rehearsal for a career designing real computers, and it's no surprise that Woz joined the first meeting of the Homebrew Computer Club. This hugely influential group, which flamed to life between 1975 and 1976, both met in the San Francisco Bay Area and shared information via newsletters. One of which included Bill Gates's famous 'Open letter to hobbyists', which lambasted enthusiasts for stealing Microsoft (Micro-Soft as it was then) BASIC software. [2]

Inspired by those meetings, Woz was already sketching out a design for a computer that he felt would be superior to the MITS Altair 8800. In fact, he had put it into practice by placing a microprocessor and memory into a video terminal that he had previously made.

Woz happily shared his schematics with other members of the Homebrew Computer Club for no fee, but where Steve Wozniak saw a chance to share knowledge, Steve Jobs saw an opportunity to make money. 'Steve Jobs came along then and saw the interest in my design,' said Wozniak in 2008 [3]. 'And he said, why don't we start a company and what we'll do is we'll make a PC board for $20 and sell it for $40 to make life easy for the people who want your computer.' Woz wasn't sure – they'd need to invest their own money to make it happen and he wasn't convinced they'd make it back – but Jobs, as we all know, was a convincing man.

Thus the Apple I was born. For Woz, though, this was an imperfect machine. He set to work on a successor. 'Within three months I designed really a computer from the ground up, the Apple II, and for some reason it wound up with half as many parts, ten times the computer. Nobody would have expected colour to be in a computer – it was just a shock to the world it could be done at an affordable cost in that year.'

This time there was no question of sharing the design for free, but the young men had a problem. They knew they could shift a thousand of the computers, but that required an upfront investment of $250,000.

The pair went to see Commodore boss Jack Tramiel to ask for the investment but could not come to an agreement; famously, Commodore went on to build the PET. The two Steves asked their friends at Atari, but the company was too busy creating the first home *Pong* game. They tried to persuade venture capitalists, but most weren't interested in investing in two young men without any business experience. Especially when one of them – Jobs – didn't at that point believe in wearing deodorant or shoes.

Their saviour was Mike Markkula, an angel investor and former Intel sales executive with enough technical nous to realise that computers were the next big thing. He invested around $80,000 of his own money in return for being one-third owner of the company – and on the proviso that Wozniak quit his job with Hewlett-Packard. After some persuasion from friends and family, and one last failed attempt to persuade the HP board that computers were more than a hobbyist niche, Woz did exactly that.

Now the hard work of producing the computer. Steve Wozniak took command of the insides, winning one crucial argument with Jobs about the need for eight expansion slots: while these ruined the elegance Jobs craved even in the late 1970s, they also allowed third parties to develop add-in cards to provide extra power or new features (such as an external modem).

With a 1MHz MOS Technology 6502 processor and 4kB of RAM, this was a powerful system for 1977. Combine this with Wozniak's genius for creating circuit boards and extracting the most possible out of the available technology, and it stood apart from any commercially available computer of the late 1970s. But its killer feature, as Wozniak alludes to above, was that it offered the opportunity to create colour graphics; in its review of the Apple II in March 1978, Byte Magazine's editor Carl Helmers chose a colour doodle of a bird created by a homebrew joystick to illustrate the piece.

Meanwhile, Jobs applied his aesthetic to the case. While the Apple II looks ordinary now, he broke away from the convention common to computers from that era of exposing their inner workings; while this may appeal to the homebrew enthusiasts, he wanted the microcomputer to break out of its soldering-iron ghetto. And he went to great lengths to ensure it worked perfectly out of the box, commissioning a new type of power supply that wouldn't require a noisy fan to keep it cool.

According to Commodore engineer Bob Yannes, the Apple II's early months were bumpy. 'The Apple II was fully assembled, but it did not have a TV output because they couldn't get around the FCC [Federal Communications Commission] emissions problems,' Yannes is quoted as saying in *Commodore: A Company On The Edge*[4]. In particular, the Apple II's strong radio interference signals meant it didn't comply with the FCC's strict Class B requirements for use of electronics in the home; this meant it couldn't be shipped with a TV output. And without a built-in monitor, this was a big problem for domestic use.

Chuck Peddle, creator of the 6502 processor, believed that Apple made three major design flaws when creating the Apple II. '[Wozniak] didn't understand the way the [6502] chipset worked,' Peddle said[5]. 'There was a guy who was hired at Apple to redesign the Apple II and make it real engineering without offending Woz.'

Steve Wozniak also supplied his own version of BASIC to run the machine, Integer BASIC, with the intention of making the Apple II easy to use – or at least, easy compared to other computers of the time. A 68-page 'mini manual' to creating programs completed the job. But Integer BASIC lacked the sophistication of

Microsoft's BASIC, leading to Markkula striking a deal to license it from the company. Thus, Apple BASIC was born.

On its release in June 1977, and with a price of $1,295, the Apple II sold primarily to hobbyists and enthusiasts. According to *Becoming Steve Jobs*, within a year Apple was shipping 500 per month [6]. It's worth asking, though, is that a big number? Bear in mind that Kit Spencer helped Commodore sell 30,000 PETs in the UK during 1978.

The truth is that the Apple II needed a push, and that push came from two things: one a piece of luck, another a piece of genius. It should be no surprise that the genius came from Woz, who developed a floppy drive for the Apple II in time for a grand unveiling at the international technology trade show CES in January 1978. The drive eventually went on sale in July for $595.

The piece of luck is that VisiCalc, the first spreadsheet program and widely considered to be the Apple II's killer app, made its way onto the Apple II at all. When Harvard student Dan Bricklin approached developers Personal Software with the bright idea for an electronic ledger, all the PETs in their office were tied up; they pointed him to the Apple II lying unused and the rest is history.

The Apple II proved to be a tougher sell in the UK. British enthusiasts had to wait until the September 1978 edition of Personal Computer World to read a review of the system, where it was listed for sale at £1,250 with 16kB of RAM – plus £35 for the carrying bag, £220 for a TV, and £15 for a cassette player. And then you had to pay VAT on top. Little wonder that the reviewers concluded: 'The Apple would be even more of a temptation were its price slightly lower.' [7]

Jack Schofield, then editor of The Guardian's computing section, remembered the Apple II fondly when interviewed in late 2019. 'It was an absolute wonder when it arrived, it was miraculous.' However, miraculous doesn't equal perfect. 'It had a 40-character, 40-column screen, but it didn't have proper lower-case letters. So you wrote everything in caps, and you did real caps as if they were inverse caps. And the machine was fairly flaky. I had two disk drives, but they were perched on top of the II, and then the monitor was perched on top of the two disk drives. So if you shoved it around, because you needed the desk space, then it could stop working. It was flaky as hell.'

The Apple II never sold in huge volumes in the UK, but was still a big influence. Instead of sitting in households around the land, its presence was felt on the future movers and shakers within the industry. Take Richard Miller, who would go on to be Vice President of Technology at Atari, CEO of VM Labs, and is now Chief Technology Officer of Pixelworks. Back in the early 1980s, he was working for a Basingstoke-based company that just happened to own an Apple II. 'Me and a couple of friends would finish work and sit down with a biryani and a Coke at 7pm and sit around writing code. We'd write this code and punch it in and then watch things happen on the screen, and create video games and things. It was just brilliant.'

Mike Fischer, the co-founder of Research Machines, is equally effusive. 'I think Woz is a genius,' he says. 'It was a partnership made in heaven. Steve Jobs wasn't an electronic engineer, but he was certainly enough of an engineer to be able to understand what the technology should be able to do and to push Wozniak to do it.' He adds: 'Even we bought an Apple II in finance until VisiCalc was available under [the operating system] CP/M.'

Just like Fischer, Jobs saw the huge potential of computers in schools, with Apple winning a crucial bid with the Minnesota Educational Computing Consortium that led to it buying 5,000 computers. This would both inject cash into the company, give Apple credibility with other school bodies, and help spread the Apple word to students, parents, and teachers alike.

The Apple II, in all its incarnations, went on to sell almost six million units over the next 16 years. It propelled Apple from an amateur outfit to a professional company, and in doing so cemented its place in the computer-maker landscape as a creator of high-end, professional computers.

While the company's next two computers, the Apple III and Apple Lisa, never found the same levels of commercial success or sheer veneration among fans, the Apple Macintosh went onto become one of the definitive products of the 1980s.

What came next

Apple kept selling the original Apple II for two years, but a number of updates – with the Apple IIe the most popular – meant its DNA carried on until the early 1990s.

Apple II Plus

Release 1979   **Price** £1,195

The Apple II Plus, or Apple II+, proved to be a hugely profitable successor to the Apple II. Shipping with the same processor but a choice of 16kB, 32kB, or 48kB of RAM, one of its best-selling expansion packs was made by Microsoft: the Z-80 SoftCard allowed users to run the increasingly popular CP/M operating system thanks to its integrated Z80 processor.

Apple II Europlus

Release 1979   **Price** £1,250

This is the Apple II that most British users will recognise. It's essentially identical to the Apple II Plus, except the video output was switched from NTSC to the European PAL standard. Unfortunately, that came with a major negative of losing colour, because it was only Wozniak's clever manipulation of the NTSC signal that allowed the Apple II to broadcast in colour. European users needed a separate video card if they wanted this luxury.

Apple III

Release 1980   **Price** £4,340

Targeted at businesses, as the price indicates, the Apple III was meant to gracefully succeed the Apple II, with enhancements such as a keyboard that produced both lower- and upper-case letters and an 80-column display. Sadly for Apple, 'succeed' wasn't an apt word, with the III beset by technical problems at launch and then torpedoed by the IBM Personal Computer.

Apple IIe

Release 1983   **Price** £1,395

A successor to the Apple II Plus, the IIe inherited the improved keyboard of the Apple III and the option of 80-column output through an add-in card. With backwards compatibility with the Apple II, and 48kB of RAM, it quickly found favour with Apple II owners as well as first-time buyers. Two years later, Apple created an 'Enhanced' version of the Apple IIe. This included a 65C02 processor and 128kB of RAM, giving it more compatibility with software being produced at the time; existing Apple IIe owners could buy an 'enhancement kit' for a suggested price of $70. The IIe's line ended with the Platinum version, featuring an almost identical keyboard to that of the Apple IIGS. The Platinum was introduced in January 1987 and kept trundling on until 1993, when Apple finally discontinued the Apple II line.

Apple IIc

Release 1984   **Price** £1,295

The 'c' here stands for compact, because this was designed as a luggable version of the Apple IIe – albeit with a new and faster 65C02 processor. Its smaller dimensions meant buyers lost expandability, but it came with five expansion cards built in and included 128kB of RAM, along with a mouse and 80-column support out of the box. It even had a built-in 5.25-inch floppy drive.

Apple IIGS

Release 1986   **Price** £999

The first sub-$1,000 Apple II series computer was also the most powerful, with a 16-bit WDC 65C816 processor that ran at 2.8MHz (notably, Apple hamstrung its speed to avoid the IIGS being faster than the Macintosh). The new chip meant it could address more memory, too, with Apple offering 256kB and 1MB options. The key letters, though, are G and S: these stood for graphics and sound, with support for 2-bit colour at 640×200 resolution, and 4-bit at 320×200. Music producers appreciated the dedicated synthesizer chip, with support for 32 voices.

Apple IIc Plus

Release 1988   **Price** £675

There was no triumphant final hurrah for the Apple II series, with the Apple IIc Plus disappointing many fans due to its lack of Plus-ness. The updated luggable included a 800kB 3.5-inch floppy drive, but with no UK keyboard the IIc Plus was never made available outside the US.

Apple IIe Card

Release 1991   **Price** £250

By 1991, the Apple II was 15 years old – yet when Apple released the Macintosh LC it recognised there was still a demand for software created for its veteran computer. With its own 65C02 processor and 256kB of RAM, the Apple IIe Card proved to be a worthy emulator for those who needed backwards compatibility.

Sources

Interviews with Mike Fischer, Richard Miller, Dick Pountain, and Jack Schofield.

1. Steve Wozniak, An Evening With Steve Wozniak, Computer History Museum, 10 December 2002
   youtu.be/rJ8IgX8RikM?t=1560
2. William Henry Gates III, An Open Letter To Hobbyists, Homebrew Computer Club newsletter, volume 2 issue 1, 31 January 1976
   digibarn.com/collections/newsletters/homebrew/V2_01/index.html
3. Steve Wozniak, Entrepreneurship and the Early Days of Apple, Haas School of Business, UC Berkeley, 22 April 2008
   youtu.be/5WBX6SACViI?t=2620
4. Brian Bagnall, *Commodore: A Company On The Edge*, Kindle edition, location 2205
5. Brian Bagnall, *Commodore: A Company On The Edge*, Kindle edition, location 2223
6. Brent Schlender, *Becoming Steve Jobs*, Kindle edition, location 939
7. John Coll and Charles Sweeten, Colour is an Apple II, Personal Computer World, September 1978, page 50

Sinclair
ZX80 and ZX81

Computers for
the mosses

In May 1979, The Financial Times bravely predicted that personal computers 'could drop to around £100 within five years' [1]. How Clive Sinclair must have chuckled. In January the following year, he announced the ZX80: a personal computer that anyone could buy for £99.95. Heck, if you were willing to assemble it yourself, it only cost £79.95.

Admittedly, this tiny computer wasn't trying to compete directly with the Commodore PET. There was no built-in monitor, the membrane keyboard was about as joyful to type on as tapioca pudding, and it had some idiosyncrasies that only a mother could love – but it also cost a seventh of the PET's price. For the first time, this was a computer that almost anyone could afford.

It also fitted perfectly into two of Clive Sinclair's key passions: a cut-down size and a cut-down price. It's how he built his first fortune, after all.

We can trace those passions all the way back to the early 1960s and the Sinclair Micro-6, a transistor-packed radio receiver that was smaller than a matchbox. 'The smallest radio set in the world,' boasted Sinclair's adverts, along with the promise that it was 'easily built in an evening!' and that 'you can even wear it like a wrist-watch' [2]. And all for 59/6, which is just shy of £3 in decimal currency. While an adjusted-for-inflation price of around £65 sounds expensive, that's a fraction of what people would pay for portable transistor radios at the time.

Together with Jim Westwood, Clive Sinclair's chief hardware engineer, Sinclair Radionics pumped out many successful products. Hi-fi amplifiers, pre-amps, a miniature FM radio kit, build-it-yourself stereos and, by the late 1960s, complete high-end hi-fi stereo units. Things were looking so good that the company placed an advert for a 'First Class Secretary' to assist the company's Managing Director [3]. In a description that would now land Sinclair in court, it specified 'an experienced, attractive girl'. At least it didn't ask for a photo and vital stats.

This stream of success – albeit punctuated with occasional flops and many delays – continued through much of the 1970s, but Sinclair's stand-out hit was the Sinclair Executive calculator in 1972. This was designed by none other than a young Christopher Curry, who would eventually abandon Sinclair and form his own company (Acorn) with Hermann Hauser.

'I had to fly out to Texas Instruments, in Texas, to get the latest chips,' says Curry. 'I flew off on New Year's Day after a party at Clive's where I was hideously drunk and had a terrible hangover. That was a dreadful flight.' But it was worth it, with Curry returning to Cambridge clutching three of the five TMS1802 prototype chips then in existence.

Curry spent the next three days soldering wires, creating a numerical keyboard using bent pieces of metal and plastic, and adding drivers for the LED display. 'About three days after I got back, I remember putting the power on and to our huge astonishment seeing numbers – slightly wrong in some cases, we got some wires wrong to start with – but seeing numbers going across the screen. That was the closest thing to magic anybody had ever seen at that time.'

The Executive would go on to make Sinclair over £2.5 million in 1974 alone with the calculator selling at £79, despite having a bill of materials that added up to £11. It was a global hit, too, with Sinclair even selling pocket calculators in Japan. On its arrival in early 1974, the Cambridge Scientific calculator proved another huge hit, selling for £49 compared to £400 for an equivalent Hewlett-Packard machine.

Sadly for Clive Sinclair, his company's run of hits dried up in the mid-1970s, to be replaced by a string of loss-making products. There was the Black Watch, aimed at electronics hobbyists, which looked great with its red LED display but suffered from a series of problems: accuracy that varied depending on the temperature; terrible battery life; and a fragile design that was so susceptible to static shocks that huge numbers were sent back for repair.

What's more, the Japanese had recovered from their early setback and were now shipping calculators into Britain at a lower price than Sinclair. Their products were more reliable too: the Cambridge had a particular problem with an oxide layer building up on its contacts due to the minimal protection offered by the membrane keyboard. Sinclair had chosen to cut costs by switching to tin-lead coating rather than gold. It was a mistake he was destined to repeat with his computers.

But the biggest drain on Sinclair's resources – or to be precise, those of Sinclair Radionics – was the prolonged and painful development of the Microvision pocket television. This dated all the way back to 1966 when Sinclair had demonstrated a prototype pocket TV at a show, although the project was shelved for several years due to production issues with the CRT. By 1975, the idea was back, with Sinclair spending tens of thousands of pounds to develop the technology.

Faced with big losses – £355,000 for the 1975/76 year on a £5.6 million turnover – and a bank that refused to extend his overdraft, Clive Sinclair needed to find a financial backer or scrap the dream of a miniature TV, along with the £500,000 he had already sunk into it. His answer came in the form of the then-Labour government's National Enterprise Board (NEB). In return for 43% of the company, it paid £650,000 into the Sinclair Radionics coffers in August 1976.

This was never going to be a match made in heaven. As part of the deal, the NEB wanted to enforce proper management structures – including a managing director to work with Sinclair – onto a company that had grown organically under the leadership of its visionary founder. Over the next two years, the NEB would continue to pour money into Sinclair Radionics even as Clive Sinclair grew more disenchanted with the partnership.

Soon after NEB had bought its stake in Sinclair Radionics, Christopher Curry 'left' to head up a separate company – first called Sinclair Instruments but soon renamed Science of Cambridge – with the idea 'to make volume products that we could sell in some of the distributor areas around the world where we had a good representation,' says Curry. Its first product? An ambitious wrist calculator for people, in Curry's words, to 'solder themselves into a mess' with.

This doesn't mean innovation dried up within Sinclair Radionics: the Sinclair Microvision TV1A Pocket Television finally went on sale in January 1977, and has even earned an entry into the V&A Collections [4]. The NEB-approved management team also invested a huge amount of money into Clive Sinclair's concept of a flat-screen television. 'He foresaw the day when the man in the street could be offered a 50in wall-mounted, flat-screen, space-saving television at a cost less than today's bulky 25in set,' wrote Rodney Dale in *The Sinclair Story* in 1985 [5].

By 1979, though, even bigger numbers were on the minds of the Radionics board. The NEB had invested almost £5 million of British taxpayers' money into a company that was now losing £100,000 per month. The management team had one get-money-quick idea: a cheap Sinclair microcomputer. But, after commissioning some initial designs, they didn't like those numbers either, with an expected development cost of £500,000 balanced against swathes of uncertainty over whether this microcomputer idea would fly. The project was put on ice.

The end was nigh for Sinclair Radionics, with the NEB breaking up the company into three different parts and bidding farewell to Clive Sinclair by means of a £10,000 golden handshake. While the NEB kept control of Sinclair's profitable instruments business, they sold off the television division to Binatone and transferred the computer project to Newbury Labs (which it also held a stake in). This project was to become the ill-fated NewBrain.

Christopher Curry, beavering away at Science of Cambridge, was also keen on computers, but his idea didn't need £500,000 of investment. 'I wanted to do the computer, not Clive,' says Curry. 'I'd been fascinated by an advert in one of the [American] magazines for something called a computer in a book. You actually bought a folder, and in one of the sheets was a PCB [printed circuit board] and the others were envelopes with bits in it. Selling it as a book instead of a computer – it was a very crude computer – seemed to be an awfully good idea."

It's around this point that Ian Williamson enters the story. 'What I could see was this incredible thirst for knowledge about microprocessors,' he told The Register in 2014 [6]. 'Certainly anyone who was an electronics engineer felt threatened by the technology and wanted to learn about it.' His idea was to create a cheap kit for engineers and hobbyists to experiment with microprocessors and program them. He created a prototype based on a National Semiconductor SC/MP evaluation board, and hooked it up to a Sinclair calculator for input and output.

Having no desire to develop the prototype into a finished product himself – he had landed a plum job with Leyland – Williamson reached out to Clive Sinclair to try to sell him the design. Sinclair introduced Williamson to Curry (neatly sidestepping Sinclair Radionics and the NEB) and for a while it seemed that Science of Cambridge would indeed develop the project. According to *The Sinclair Story*, there was even an agreement drawn up for a '£5,000 down payment plus royalties' [7].

A phone call to National Semiconductor – usually shortened to Nat Semi – would ultimately change the course of history, however. On hearing that Curry was willing to buy in bulk, an enterprising sales manager suggested that Nat Semi could help with the logic design. It was an opportunity too good to pass up and ultimately became the MK14: Curry assembled and taped out the PCBs, then used the notes created by the Nat Semi sales engineer for the logic design. That same engineer also helped with the early programming work, which was finished off by David Johnson-Davies.

This project became the £40 MK14, a microprocessor kit comprised of 14 pieces. While it was released under the Science of Cambridge banner, it already showed all the hallmarks of a Sinclair computer, right down to a cheap membrane keyboard and long delays for delivery: a reviewer in the May 1979 edition of Practical Electronics claiming that he 'received one in January which was ordered in the previous summer! Others have experienced waits from two to four months'. [8]

Still, the review concluded with the words, 'my firm opinion is that no electronics enthusiast or engineer should be without one in today's technology' and estimated sales figures of 20,000 suggest others agreed. The MK14 proved beyond doubt that there was a market for microcomputers; if the MK14, which was essentially a My First Microprocessor training kit, could sell in such volume, what was the limit for a 'proper' microcomputer?

It was a thought that prompted Curry to set up his own company with Hermann Hauser, which would lead directly to the Acorn Microcomputer System 1 and, in a more roundabout fashion, to the BBC Micro. And the moment Clive Sinclair was released from the shackles of the National Enterprise Board, he wasted no time in instructing Jim Westwood to get to work and produce a cheap Sinclair microcomputer.

Westwood set to the task with his usual enthusiasm. 'It's a challenge managing to achieve something without using expensive components and I like that challenge,' he told Sinclair User in 1982 [9]. 'Of all the products with which I have been involved I think the ZX80 is my favourite. It was a real breakthrough in the use of cheap components. It is something which ought to be in the Ark by now but I am still proud of it.'

As well he should: from a 'mess which we call a breadboard' he created a computer that sold around 50,000 units and gave many Britons their first taste of a personal computer. Admittedly, he did so using readily available components, with the only Sinclair-created ingredient being the firmware. Few cared: to include a 3.25MHz Z80 processor in a ready-made computer that cost less than £100 was quite a feat. (Well, a replica Z80 processor: it was actually made by NEC.)

There were inevitable sacrifices, with the biggest being the 1kB of RAM and 4kB of ROM. Having only 1kB of memory severely limited the ZX80's capabilities, to the point where it was a machine to experiment with programming on but little else. You could load programs from a tape recorder, but if buyers were hoping for a singing and dancing computer then they were to be disappointed. If the ZX80 ever entered a Celebrity Computers Come Dancing contest, it would be out after the first show.

It's debatable whether it would have won points for style: its tiny vacuum-formed plastic case lent it a futuristic air, but still felt cheap. And those output grilles at the top were not what they seemed. 'What always amused me was the printed-on cooling slots,' says Jack Lang, co-founder of the Raspberry Pi Foundation and a key part of Cambridge University's hugely influential Computer Laboratory.

If people became frustrated by the 1kB of RAM, there was an expansion port on the rear. Memory upgrades were initially criminally expensive – £300 for 16kB – but the good news for early investors is that prices did come down. By 1982, you could buy a third-party 16kB add-on pack for £32.95.

The ZX80 was also the first outing for Sinclair BASIC, created by John Grant from Nine Tiles. His skill was to make it easy for users to pick up BASIC through on-screen prompts that would tell them if they were making a mistake, and to make things that little bit easier you could press, say, 'Q' and 'PRINT' would appear on screen. As Grant explained in an interview on the Floppy Days podcast in 2018 [10], this shorthand also meant the BASIC took up less space as, 'I didn't then have to have code that recognised "P-R-I-N-T" and translated that into the keyword PRINT'. It was tricks such as this that enabled Grant to perform the minor miracle of squeezing a beginner-friendly version of BASIC into 4kB of ROM. To make life even easier for newcomers to BASIC, not only would you be told of a syntax error during typing but Sinclair included a 128-page manual that introduced newcomers to the language.

Not that the ZX80's simplicity stopped Clive Sinclair from giving it some pizzazz at its launch in January 1980. 'It's the biggest leap forward we've ever made in terms of price and technology,' he claimed [11]. Adverts, meanwhile, led with the boast that: 'Inside a day, you'll be talking to it like an old friend!'

What Sinclair didn't go into were the little irritations that bugged early users. The biggest of these was screen flicker: this happened whenever you typed something in. While Clive Sinclair defended this at launch with the claim that it was a feature, because it gave the user 'positive feedback' [12], it was actually a cost-cutting measure: the Z80 processor controlled the output to the screen, rather than using a dedicated video chip.

With such limited RAM, programmers also needed to compromise on either program complexity or output: for instance, using the ZX80's maximum output of 32 characters per line and 24 lines per page consumed over 600 bytes. Or viewed the other way, if you were running a program that consumed over 900 bytes then you would be left with only a handful of lines for output. As for sound... well, silence is golden.

We should also emphasise that the ZX80 (and the ZX81) could only output in black and white, with Sinclair fans waiting until the Spectrum for glorious colour. It used the same trick as the Commodore PET for creating 'graphics' too: with no hardware sprites, it relied on block-based graphics that were part of the character set. Early games writers needed to come up with imaginative innovations to ensure their games looked good.

Then there was the touch-sensitive keyboard with no moving keys. Again, Sinclair's spin machine was working at full speed to emphasise the fact it was spill-proof, and that you could even pour a cup of coffee over the ZX80 without ill effect, but there was soon a booming market in third-party keyboards due to the fact that it had to be used with extreme care to avoid brushing the wrong letters.

In his review for Personal Computer World [13], David Tebbutt memorably wrote: 'Typing gives a sensation of drumming your fingers rather than of doing anything useful.' But he quickly went on to say: 'This is a totally mistaken impression because it really works rather well.'

Indeed, Tebbutt was one of many in the British press to lavish praise on Sinclair's computer despite its drawbacks. 'The ZX80 appears to be a well thought out machine both in terms of its hardware and software,' he wrote, before concluding: 'I hope Mr Sinclair and his merry men of Cambridge can cope with the expected flood of orders.'

Cambridge was also the start of Sinclair's invasion into America with the ZX80, although this time it was Cambridge, Massachusetts, which sits across the river from Boston. 'I met up with [Clive] there on his way back from Vegas,' recalls Nigel Searle. 'He had his ZX80 prototype with him, which he hooked up to the TV in his room at the hotel. By lunchtime the next day, we had rented an office in Boston, incorporated the business, and I rented an apartment right next to the office. Then we went and had lunch.'

The great benefit of exporting the ZX80 to the US market was that it needed minimal modifications, other than improvements to its RF radiation output. And the US public lapped it up. 'We ran some audacious ads,' says Searle. 'We were just getting such a phenomenal response, we would run an ad that cost a couple of thousand dollars and get $20,000 or $30,000 worth of sales.'

Even as Sinclair Research was ramping up production of the ZX80, Jim Westwood was focusing his effort on the ZX81. Rather than create a more advanced computer, his brief was to bring the price down yet further. Westwood achieved this feat by reducing the number of chips from 21 to 4, thanks to an Uncommitted Logic Array (ULA) made by Ferranti (a company that crops up often in the history of 1980s British computers, and not always positively).

Aside from this, the ZX81's fundamental specification was unchanged from the ZX80: still 1kB of RAM, still a Z80 processor. But there were numerous improvements. First, the screen flicker was gone. The cunning trick was a SLOW mode: here, the Z80 would dedicate a quarter of its time to running the current program and the rest to display output.

Thanks to an extra 4kB of ROM, Steve Vickers from Nine Tiles made some neat enhancements to the BASIC, most notably adding support for non-integers and over 30 extra functions. This brought it much closer to the 'official' ANSI standard for Full BASIC. Richard Altwasser is fulsome in his praise for what Vickers achieved. 'He pretty much single-handedly wrote, tested, and debugged the Spectrum BASIC – and wrote the instruction manual.'

With the help of vacuum forming, the hardware was also improved. Legendary industrial designer Rick Dickinson overhauled the design, and would win a 1981 Design Council award for his efforts, even if the controversial membrane keyboard remained in place. 'The ZX81 looks and feels good,' wrote Tim Hartnell in Your Computer Magazine [14]. 'It is about the size – 7in by 7in – and weight of a paperback book, finished in matt black, with a matt-plastic keyboard finished in red and black.'

While castigating Sinclair for the instability of his test machine, and the 16kB RAM expansion pack in particular, Hartnell still described the ZX81 as a 'very good first computer'. He concluded: 'You will learn a good deal, have considerable fun, and when – in eight months or so – you are ready to move on to another machine, you will have enough knowledge to know exactly which computer to buy.'

David Tebbutt was even more unequivocal in his praise. 'He's done it again,' he wrote in the June 1981 edition of Personal Computer World[15]. 'Uncle Clive has come up with a lovely product which will have enormous appeal to people wanting to find out more about computers, but without it costing an arm and a leg.'

And the initial price was very low indeed. £49.95 if bought as a kit and £69.95 assembled was difficult to argue with, and unlike the ZX80, Sinclair wasn't charging an additional £8.95 for the mains adapter (although postage and packing was still an extra £2.95). By this time, though, Sinclair knew there was a lot of money to be made through extras. At launch, it was selling the 16kB RAM pack for £49.95, and this would soon be followed by a printer for the same price.

The usual Sinclair advertising blitz soon followed, with ads in national newspapers promising[16], 'Inside a day, you'll be talking to it like a new friend.' Note the subtle change from 'old friend' in the ZX80 ads to 'new friend', an early Easter egg for true Sinclair fans. The ad also declared that 'you'll find it of immense practical value. The computer understanding it gives you will be useful in any business or professional sphere. And the grounding it gives your children will equip them for the rest of their lives.'

The 'free course in computing' promised by the same advertisement came via a ring-bound 212-page guide, adding yet more weight to the feeling of value. Little wonder, then, that it proved a huge success. Production at the Timex factory in Dundee, where the ZX81 machines were assembled, rose from 10,000 per month at launch to 60,000 per month a year later, allowing Sinclair to ship yet more to the US market.

Life was good. So good that Sinclair Research was determined to double down on its advantage and produce the UK's biggest selling home computer of all time: the ZX Spectrum.

ZX81: What was it good for?

One of the most fascinating things about browsing through back issues of Sinclair User is to discover how people used their ZX81 computers. In the July 1982 edition [17], Julian Moss explained how it was being used by amateur radio enthusiasts. For example, in aerial design: 'the computer can be used to work out the dimensions of an aerial for a particular frequency, and even to calculate its theoretical performance'.

In August 1983, we discovered how a ZX81 – with the help of a 64kB memory expansion pack – was being used to handle the payroll of a small business. 'I used to do it all in my head, with the help of a ready reckoner and it took me a day and a half,' said Vera Sims [18]. 'Now my husband has written a program which does it for me and it takes a few hours.'

Two months later Robert Entwhistle explained how he had hacked the ZX81 to support two keyboards and play music, courtesy of an I/O port connected to a loudspeaker. Admittedly it helped that Entwhistle was an engineer. 'When [Entwhistle] wanted to increase capacity from 1K to 2K,' wrote Claudia Cooke [19], 'he fitted the new chip.'

With a bit of imagination, some extra memory and quite a lot of expertise, it turned out there was very little the ZX81 couldn't do.

The ZX81 legacy

While the ZX80 never truly moved beyond toy computer status, the ZX81 was a key spark behind the explosion of British children learning to program. 'It was the first computer I programmed,' wrote Julian Tysoe in response to our online survey asking people about their most-loved computers. He went on to get a degree in computer science and spend 27 years 'programming MUMPS/M/CACHE, which is about as close to ZX BASIC as you can get in the professional world!'

It's a similar story for Ian Wilson, who learned machine code programming on the ZX81. Has it had an impact on his career? 'Yes, I've worked in IT all my life. The early games grabbed my attention and I'm still an avid gamer today in my 50s.'

Geoff Airey, like several others who took part in our survey, saw the ZX81 as a stepping stone computer. 'It opened my eyes to more powerful computers later,' he wrote. 'They weren't just something to play games on like the Atari 2600, you could make it do what you wanted.'

We'll leave the final words to John Parkinson. 'My first computer: slow, infuriating, magical.'

Sources
Interviews with Christopher Curry, Hermann Hauser, Jack Lang, and Nigel Searle.

1. Rodney Dale, *The Sinclair Story*, Gerald Duckworth & Co Ltd 1985, page 95
2. Sinclair advertisement, Practical Wireless, February 1965, page 1004

3. Rodney Dale, *The Sinclair Story*, Gerald Duckworth & Co Ltd 1985, page 29

4. V&A Collections, Microvision TV1A Pocket Television
   collections.vam.ac.uk/item/O1204914/microvision-tv1a-pocket-television-television-pemberton-john/

5. Rodney Dale, *The Sinclair Story*, Gerald Duckworth & Co Ltd 1985, page 87

6. Tony Smith, Ian Williamson: *The Engineer who gave Sinclair his first micro*, The Register,
   theregister.co.uk/2014/01/16/archaeologic_ian_williamson

7. Rodney Dale, *The Sinclair Story*, Gerald Duckworth & Co Ltd 1985, page 91

8. A.A.Berk, MK14 review, Practical Electronics, May 1979, page 50

9. Claudia Cooke, *Hiding his light behind Sinclair*, Sinclair User, November 1982, page 26

10. Floppy Days, episode 85: interview with John Grant, 8 July 2018,
    floppydays.libsyn.com/floppy-days-85-interview-with-john-grant-developer-of-zx80-os-basic

11. Naunton Pugh, *"Computer 'for all the family' is launched"*, Cambridge Evening News, Tuesday 29 January 1980

12. Peter Large, *Clive Sinclair does it again – but will the ZX80 be a winner?*, The Guardian (London),
    Wednesday 30 January 1980, page 15

13. David Tebbutt, Benchtest: Sinclair ZX80, Personal Computer World, April 1980, page 55

14. Tim Hartnell, Review: The Sinclair ZX81, Your Computer Magazine, June 1981, page 12

15. David Tebbutt, Benchtest: Sinclair ZX81, Personal Computer World, June 1981, page 66

16. The Guardian (London), Tuesday 24 March 1981, page 8

17. Julian Moss, *Radio Sinclair*, Sinclair User, July 1982, page 22

18. Claudia Cooke, *Taking the strain out of calculating wages*, Sinclair User, August 1983, page 78

19. Claudia Cooke, *Getting the ZX81 to do things others cannot*, Sinclair User, October 1983, page 84

Sinclair ZX80 and ZX81

Commodore VIC-20

The ice-breaking
computer that wanted
to be your friend

For a computer designed to be cheap and cheerful, the VIC-20 had many fathers. The MOS Technology team, led by Al Charpentier, who created a graphics chip that no one appeared to want. Robert Yannes, who designed a VIC-20 prototype based around that chip in the space of three days. Robert Russell, who took that prototype and turned it into a computer (with a lot of help). Michael Tomczyk, who passionately battled to ensure the VIC-20 project happened. A brilliant set of Japanese engineers who created the final design.

And the Sinclair ZX80.

The ZX80 enters the story because it had just been released in the UK for £99.95 when Jack Tramiel, the Commodore boss, visited London. 'The Sinclair was directly responsible for Commodore doing the VIC-20,' said Chuck Peddle [1], creator of the Commodore PET. 'Fundamentally, Jack always was into the market and the market was buying the Sinclair machine.'

This was bad news for Peddle. He, along with all the other senior management within Commodore, had convened for a meeting in Burnham Beeches Hotel, just outside Slough, in April 1980. Over the course of a week, they would plan the company's priorities for the coming year. For the first half of the first day there seemed one clear answer: a high-end, colour update to the PET that would usurp the Apple II. This colour PET was the computer that Chuck Peddle, then effectively in charge of Commodore's computer division, was pushing. He even had a prototype to show off. According to reports, the general managers of each country agreed. And then Jack Tramiel walked in.

While some accounts of this seminal moment in Commodore's history pump up the hype, making it sound as if Tramiel threw a literal grenade into the middle of the room, Michael Tomczyk, newly appointed as Tramiel's Assistant to the President and Marketing Strategist, remembers it less dramatically. 'He stood up and said, "Look, I would like to have a low and inexpensive colour computer instead of this one [pointing to the colour PET prototype]. We can still do this one, but I want the introductory computer first." And Chuck Peddle was horrified because he thought he was going to just get a rubber stamp of approval for the colour PET.'

At this point, Tramiel left the room, leaving confusion in his wake. 'Many of the seasoned executives felt that a little colour computer for a few hundred dollars was not going to be worth all the trouble and it wouldn't bring in enough money when we could easily sell the colour version [of the PET] for $1,000 or in that range,' says Tomczyk.

'They were just thinking in terms of dollar signs,' he adds. 'I was thinking in terms of "hey, if we can get people hooked on a small, cheap colour computer that they can hook up to their TV set, they will naturally go to school and see the PET in schools – we were dominant in schools – and then when they graduate school, they'll buy our business system. But we've got to get them hooked on the small system first. It's got to have some games and something to really magnet people into it. And I made that case very strongly.'

Tomczyk describes a 'triumvirate' who backed the VIC-20. First, the Brits of Kit Spencer and Bob Gleadlow (then head of Commodore UK), then the Japanese duo of Tony Tokai, general manager of Commodore in Japan and his technical guru Yashi Terakura, and finally – to a lesser extent – Harald Speyer, general manager of Germany. Together, these were the dominant markets for Commodore's PET sales: in the USA, the PET had fallen behind both the Apple II and Radio Shack TRS-80 despite being the first to market. While it had been a tactical decision by Tramiel to focus the computer group on Europe and Japan, he was now keen to resume battle on the home front and saw the low-priced computer as his weapon.

Others were less convinced, leading to some lively discussion over the remainder of the day. Not that Tramiel minded conflict between his senior team: part of his management style was to actively pitch teams against one another – as would later happen in the VIC-20's development. 'I think there may have been a bit of friction,' says the unflappable Kit Spencer, then in charge of marketing for the UK, 'because Chuck wanted to move more upmarket than downmarket. And he considered the VIC-20 to be going downmarket from the PET.'

For Spencer, though, this wasn't an either/or situation. 'Look at what IBM did. Going into the traditional computer market was massive, but so was the home market. It's a question of finance and priorities. Basically, I said, both are good. I'm quite for the home market. I can see it coming. We need to do something. But that doesn't mean we abandon what we've got and build on that either.'

Tramiel returned the following day. 'He said, "OK, so what do you all think?" All of us went around the table, and we all made our case,' says Tomczyk. 'After we finished, Jack stood up, pounded his fist on the table, everybody fell silent, and he said, "Gentlemen, the Japanese are coming. So we will become the Japanese." And everybody knew what that meant. If we didn't do the small computer, because there really wasn't anything like it, the Japanese would just come in and they would take the market and everything would happen the way I described it. People would buy the small computer, and then buy the next higher one and the next higher one, the Japanese would have all that and they would just take the market away from us if we didn't do it ourselves. And Jack didn't have to explain that very much. He just said that one sentence and that almost ended the whole day.'

It's worth pausing a moment here to consider the force of nature that was Jack Tramiel (he died at the age of 83 in 2012). Born in Poland, he and his parents were forced into the Jewish ghetto in Lodz when the Germans invaded in 1939, eventually being sent to Auschwitz. Only he survived, in part because he went to work in a labour camp rather than Auschwitz itself. He emigrated to the US in 1947 and went on to join the army, before creating a successful company – Commodore Business Machines – around typewriters first and then calculators.

Notably, and perhaps amazingly, he wasn't bitter towards Germans for the Holocaust. Tomczyk, who grew close to Tramiel during their time working together, asked his boss why that was. 'He told me once that it wasn't the Germans that killed Jews, it was the rules that killed the Jews and the Germans always follow rules. So he really said, "I have nothing against the German people, but you know, we have to prevent this from happening ever again."'

It's no coincidence that immediately after the conference at Burnham Beeches, Tramiel flew to Germany to negotiate the purchase of a struggling electronics plant. 'Jack met with these German officials,' says Tomczyk, 'and in the meeting one of the officials said, "Why should we do this?" And Jack looked at them and said because you owe it to me. But also this would be terrific PR: an Auschwitz survivor comes back and does business in Germany. I would like to get some benefits from the government and have all the bureaucratic nonsense taken care of, and then I will save that company from going bankrupt and we'll start making Commodore computers here for Europe." And the Germans looked at him and said, "Well, that sounds very reasonable."'

When Tomczyk later asked how Tramiel deals with the Holocaust, the Commodore boss had a simple answer. 'He said, "I live in the future." You know, without batting an eye,' says Tomczyk.

On their return to Commodore's Palo Alto office, Tomczyk set to work on what he describes as 'a very quick memo' about what was needed to make the new computer work, but reportedly ran to 30 pages. 'When I was done, I looked at it. And I'm not a formal person. So I drew a large happy face with a beard and moustache, which has become my logo and my trademark ever since.'

While his official job title was Assistant to the President and Marketing Strategist, Tramiel had hired Tomczyk on the basis he would spend the first six months watching what went on and doing very little. He could visit whatever corner of the company he wished, speak to whoever he liked, and even barge into Tramiel's office at any time without needing to knock. Tramiel knew there was a place in the company for this enthusiastic marketing man, who had given up a job as a general manager of a small special effects company in San Francisco to join the microcomputer gold rush. It was just a matter of finding out what job he should do.

Tomczyk took advantage of the 'barge in whenever you like' clause of his contract and threw the 30-page memo onto Tramiel's glass desk. 'He said, "What's that?"' recalls Tomczyk. 'By the way, he had a deep booming voice. He was short, rotund, bald, and he could make frightening expressions with his face. And he was very, very intense and intimidating when he wanted to be. But anyway, he looked at me and he said, "So what's that?" And I said, that's everything that should be done with a new computer. Make sure whoever's in charge does all these things. And so about a week later, he came back into my office. He threw it on my desk, and I said, "What's that?" And he said, "That's everything should be done with the new computer, you're in charge of making it happen."'

This presented some challenges. Not only was Tomczyk new to Commodore, he was new to the computing industry. What's more, he would be instructing people technically superior to him, who knew far more about microprocessors and engineering, and over whom he had no authority. But he also had two things going for him: a persuasive nature and Jack Tramiel's backing. In Commodore, where job titles were less important than being 'in' with Jack, the latter was a huge advantage.

What Tomczyk didn't know, along with everybody else based in Commodore's west coast headquarters, was that a young engineer working at MOS Technology, the Pennsylvania-based semiconductor manufacturing company Jack Tramiel had bought four years earlier and which directly led to Commodore creating the PET, had also been inspired by the ZX80. His inspiration, though, was based around anger.

To understand that anger, we need to rewind a year to 1979. The engineer in question is Robert Yannes, who was a big fan of a cheap video chip created by one of Commodore's most influential engineers, Al Charpentier. Charpentier had started working on the chip in late 1976, almost immediately after MOS Technology had been taken over by Commodore. MOS already made video chips for Atari, which were embedded into the games cartridges along with the software. Charpentier sensibly thought this was a waste: why not include a more powerful video chip to sit alongside the microprocessor? He set to work, and finally finished the chip a year later.

It was an impressive piece of technology for the time, able to display black, white, red, cyan, purple, green, blue, and yellow. With support for light guns and bitmapped graphics, it seemed perfectly tuned for gaming systems. There was just one problem. Atari, then the most obvious customer for what Charpentier dubbed the Video Interface Chip, or VIC, wasn't interested. Unbeknownst to him, it was already developing a more powerful video chip in-house.

Commodore publicly demoed the VIC at the January 1978 Consumer Electronics Show (CES), with the idea of selling it direct to other computer and console makers, but no one was interested. The big problem? Where the PET could display 40 characters per line, the VIC only showed 22. Even Steve Wozniak was impressed by the chip's colour performance in the on-stand demo, but the character limit made it seem babyish in comparison with the Apple II.

There was also limited appetite for MOS Technology's chip within Commodore. Paper concepts of a computer based on the VIC were drawn up during 1978, but with the company focused on producing a more advanced, 80-column PET it fell to the back of the queue.

In November 1978, the VIC's fate finally took a favourable turn when Charpentier met Yannes. Robert Yannes was then studying at a small university in Pennsylvania, near the MOS Technology building, and during an interview with Charpentier he noticed the VIC chip. Enthused by its capabilities, Yannes asked if he could build his senior year project around it. Charpentier gladly agreed.

In July 1979, Charpentier hired Yannes, who grew just as frustrated as his boss about the VIC's lack of success. 'I thought this was a great chip and hardly anyone seemed interested in it,' Yannes is quoted as saying in Brian Bagnall's definitive history, *Commodore: A Company On The Edge* [2]. 'My job was to figure out how to get people interested in using the VIC chip.'

In the end, he needed just one person to be interested: Jack Tramiel. Even more crucially, the timing had to be right. By May 1980, the Commodore boss had endured a number of frustrating months as he waited for his engineers to produce the low-cost computer he had demanded at the Burnham Beeches meeting.

Meanwhile, Yannes had seen the praise lavished on the Sinclair ZX80 but considered it 'just terrible'. 'Here's this crummy little Sinclair computer out there for 250 bucks, and we can do a real cool one for just 200 bucks,' he said. 'To hear all this good press about people being excited about it, I just said, "This is crazy. We have this little video chip here that can make a great little colour graphics computer for less than the Sinclair. We can even put a real keyboard on it and make a nice product instead of that horrid thing."'

In the space of three days, and using all the knowledge he had gained from his senior year project, Yannes put together a basic wire-wrapped prototype based on the KIM-1 board (the precursor to the Commodore PET), built a case using sheet plastic, and stuck on a PET-style calculator keyboard. He now had a 22-column colour computer that worked if you hooked it up to a TV set.

Not only was his boss impressed, so were the senior managers at MOS Technology. They knew Tramiel was due to visit and was hunting for a low-cost computer. This was a chance to score a coup over the west coast team led by Chuck Peddle, who would also be at the meeting. They set to work turning Yannes's tinpot prototype into something more professional, complete with a moulded case. What they couldn't do was load BASIC or any sort of operating system, so Yannes programmed a demo instead.

While Yannes wasn't present at the meeting with Tramiel, he ensured the presentation would be a success by using the demo to highlight all the features of the VIC he had learned to love. Along with improved colour capability since its initial design, it also included advanced audio. In his demo, screens slid in and out of view, from top to bottom, from left to right, each extolling the VIC's virtues and accompanied by music. 'I kept the whole demonstration in black and white, and then the very last thing said, "Oh, and by the way, it has colour."'

Tramiel was convinced. So convinced he commanded the MOS Technology team to turn the concept into a finished computer, and that meant showing a fully formed prototype at the June 1980 CES. Which was just two weeks away. While Charpentier and Yannes set to work making this happen, an annoyed Peddle decided to focus his west coast team on producing their own prototype based on the VIC chip. The crucial difference: theirs would include BASIC.

Bill Seiler described their prototype as a 'G-job'. These hacked-together designs, usually made in garages (thus the 'G'), were proof-of-concepts that would never be the foundation of a production design – but they would work. Their version of the computer was literally hacked together, including a sawn-off PET motherboard with the VIC graphics chip attached by wires. They then stuffed the electronics into an old calculator case that Seiler grabbed from a dumpster, before squeezing in a calculator-style PET keyboard.

Now to get software onto it. While the hacked-apart board included a cassette port, the VIC's video system was different to the PET's. Time for Peddle and his team to put in a couple of all-nighters to make it work, which they managed with a day to spare – just time enough for them to convert some PET games to work on the 22-column screen of their new computer.

Tramiel used CES to pitch the two prototypes against one another. A privileged few were invited into the VIC-20 demonstration room, where they could see the clumsy-looking but fully working G-job on one side, and the more glitzy but basic prototype created by Yannes on the other side. In truth, calling it a prototype is generous: it was less a computer, more a demonstration for the VIC chip.

Unfortunately for Peddle, the cards were stacked against his team because he was still in Tramiel's bad books. He felt that his chief engineer had let him down because he didn't believe in the low-cost computer, while also failing to deliver the Apple II killer for business. In a move that was typical of Commodore's bullish founder, he decided to in effect replace Peddle as head engineer, bringing in Tom Hong from Apple and instructing him to build a production version of the east coast's VIC computer within six weeks.

Looking back on the process through the prism of 40 years, it seems miraculous that the VIC-20 was produced at all. Not only was Hong placed in an almost-impossible position – he left after three months – but so was his allocated engineer, Robert Russell. It didn't help that Hong decided to base the VIC-20 on the Yannes prototype rather than the west coast's working, if bodged, machine.

It's hard not to feel sympathy for Russell when reading his description of those weeks. 'They were kind of like, "OK, Bob, take this hardware and put software on it,"' he recalled [3]. 'It was like, "Argh! You've got to be kidding me.' It had none of the peripheral ports figured out and not enough ROM to sneeze in.'

Not only was Russell inexperienced at hardware design – he was essentially a software engineer – but he lacked resources to call upon. MOS Technology felt that its job was done while Peddle's team was now focused on creating the Apple II rival. 'They didn't want to have anything to do with the VIC project,' said Russell.

Nonetheless, Russell set to work. Naming the new computer 'Vixen', he would work at the main Commodore office during the day and then set off to Moorpark – where Peddle's design team were based, and where the high-quality equipment was to be found – to work during the night. While he could do marvels with the software, he needed the help of Ed Seiler to develop the hardware. Tricky when Seiler, along with the rest of Peddle's team, was meant to be working on their own jobs.

Fortunately for Russell, he had the Jack card in his back pocket. 'I conned Seiler into helping me rebuild stuff and design things like the memory add-in board, so you could actually have enough memory to do something with it,' said Russell. He would also play his Jack card to persuade other members of Seiler's team to help him hit his six-week target, to the huge frustration of Peddle. It was like being told you had lost a game, but now you had to help the winning team claim their prize – and that team had already gone off to celebrate.

Nevertheless, they played a huge part in bringing the Vixen to market, whether that was adapting BASIC or producing colour text. Seiler also designed a cartridge port, whose sheer size (needed because the port would also be used to expand its memory) led to the huge cartridges that VIC-20 users will still remember.

While the VIC chip could produce graphics up to 192×200 pixels, there was no guarantee that the TV screens to be used with the VIC-20 would support that resolution. So the team came up with a clever solution: they would add a border of varying size around a 176×184-pixel rectangle. Users could even control the colour of that border.

To Peddle's undoubted frustration, yet more features from the rejected G-job prototype made it into the final unit, including their adapted version of BASIC. And to add salt into that gaping wound, time constraints led Seiler's team to implement many features of the Apple II rival they were meant to be working on into the VIC-20. Little wonder that Peddle soon left Commodore to start his own company, Sirius, that would go on to create the popular Victor computer.

Even as the hardware was being developed, Michael Tomczyk was embracing his unofficial role of VIC-20 Czar. His guiding principle was simple: 'It has to be user-friendly. User-friendly was a term that was just coming into vogue, and I embraced it and made it my slogan. And then Chuck Peddle left the company and took some engineers with him, and even before he left those engineers didn't want to work on the new computer.'

So Tramiel took a decision: hand Russell's almost-finished prototype to his Japanese design team, led by Tony Tokai and Yash Terakura, and let them produce a version ready for production. To avoid confusion, the only person from the US office allowed to liaise with them was Tomczyk. 'That was tough to coordinate,' he admits. 'I'm not an engineer, I didn't sit where the designers were, but I had to make sure that the features I wanted were included.'

Some of those features proved unrealistic due to the restrictions of the VIC-20's design, but he kept to his user-friendly mantra. 'One example is I wanted it to build in an RS-232 interface so we can connect it to a telephone,' says Tomczyk. (A year later, Commodore would release a $99 modem – commissioned by Tomczyk – that would help to make millions of dollars.) 'We wanted to make sure we have full-size typewriter keys.'

Tomczyk flew over to Japan to check on the progress of Tokai and Terakura, as they worked on the final production model of the VIC. Whilst there, Tokai took him to a Tokyo electronics store where NEC happened to be showing off an early design of one of its unreleased computers. 'It was beautifully designed, kind of like an Apple II, but with bright orange function keys,' says Tomczyk. 'I went crazy when I saw those function keys. I said, if we drop this into the software community, they will be able to assign those keys to do all kinds of things related to their software.

And they will be adaptable and flexible and valuable. So we must have those function keys.'

It was time to get serious about names. Having rejected the name Vixen because of its 'sexy woman' connotations, Tomczyk had settled on Commodore Spirit. 'Well, Tony Takai called me on the phone and said Michael-san, you cannot call it the Commodore Spirit. And I said why? And he said, 'In Japan, the word spirit doesn't mean like Caspar the Friendly Ghost. It means flesh-eating, soul-feeding ghoul from hell.'

With his spirit crushed, Tomczyk went back to the VIC name that was already popular with the original engineers. He explains his reasoning like this: 'Vic sounds like a truck driver's name, so I'm going to add a number. So 20 is a friendly number. It has 22 columns and 20 is pretty close. So we'll call it the VIC-20. Tony said OK, but I'm going to call it something else in Japan.' Tokai eventually settled on the VIC-1001, due to the popularity of the film *2001: A Space Odyssey*.

Tomczyk flew out to Tokyo for the launch, which wasn't at a trade show but a department store. 'We had a couple of tabletop booths, like chest-high shelves, where the VIC-20 was,' says Tomczyk. 'Tony, Yash, and myself would be standing, talking constantly about the crowd reaction, what the features were, what should be done next, and Japanese engineers from other companies kept sneaking into the booth with screwdrivers trying to take apart the case to see what we had done there. We had to keep shooing them away constantly.'

With the Japanese version of the computer finished, it was time to ship units in the rest of the world, but there were still a couple of, literally, key decisions to be made. 'The engineers came to me and said, we've got one key that's not assigned [the Yen key]. Would you like to assign something to it, Michael? And I said, "Well, yeah, let's put an English pound sign on to help Kit Spencer, because then we won't have to make a new English keyboard. You can just use the US keyboard and that will get this to the UK faster.'

At this time, Europe was still outselling the US – Spencer would later joke that he was taking a demotion by moving from his Swiss base to become head of marketing for Commodore US – and without the enviable distribution of Tandy's computers, or Apple's far-reaching tentacles, it had a challenge. While the wisdom of hindsight makes it obvious that the VIC-20 would go on to sell a million units, the computer's success looked far less assured at its launch in early 1981.

Fortunately, there were many at Commodore determined for it to succeed, and Jack Tramiel had given Michael Tomczyk the authority to hire people to make it happen. His chosen route: the VIC Commandos.

'I hired these young guys in their 20s – one was as young as 18 – and they were all self-taught programmers,' Tomczyk said. 'They took some of the old black-and-white Commodore [PET] games that we didn't have to pay royalties for and turned them into colour versions for the VIC-20. And we did a six-pack of games including one that I called the *Blue Meanies From Outer Space*, which I think is a Beatles lyric. Then I wanted a six-pack of productivity software – home mortgage, home budget, personal finance – to show the utility aspect of the computer.'

Neil Harris proved to be one of Tomczyk's most inspired hires, rapidly rising within Commodore's ranks between joining in February 1981 and leaving in summer 1984 – to join Jack Tramiel's new adventure with the Atari ST. But that was in the future. When Harris started at Commodore, his main experiences had been selling computers – both the PET and the Apple II – and teaching BASIC programming courses. He'd also written a few articles for magazines. In short, he had an unusual and valuable mix of skills. Little wonder that Tomczyk hired him the day after they met at a Commodore job fair.

Harris's first task? Salvage the VIC-20's user manual. 'It had been commissioned to an outside company that wrote a very fluffy manual that didn't have a lot of meat in terms of actual useful tutorials about how to use the computer,' says Harris. 'Mike showed me the draft of the manual, and said, "You know, this is terrible. We need to fix this." And I said, "I'm your man. Where's my word processor?"'

But there was no word processor because the PET that Tomczyk had requisitioned still hadn't arrived. This reinforced what would soon become clear to Harris: anyone working on the VIC project had zero status at Commodore. This was in stark contrast to the colleagues they shared their small sales office with in a Philadelphia suburb. 'They were the business PET computer guys and we were the little toy computer guys,' says Harris. 'But we were coming in early and working late. The other guys would leave at 5 o'clock and if we didn't have a disk drive and a cable, we would suddenly end up with a disk drive and a cable and the other guys would find that there's had gone missing and they could requisition a new one. Everybody was happy.'

While they could get away with stealing the odd cable, a whole computer was a different matter. This meant Harris had the interesting challenge of writing a manual without a word processor, so he dug out his typewriter and started tapping. Within two weeks, Tomczyk had a first draft of the VIC-20's manual in his hands, complete with example BASIC programs, written by Harris, to show new users exactly what they could do. With the addition of some appendices from others in the Commando team, this comprehensive and user-friendly manual would go on to be read by over a million VIC-20 buyers.

Harris would also contribute two games to the six-pack of tapes – *Super Slither*, based on an arcade game of the time and familiar to anyone who has played *Snake* on an old Nokia phone, and a version of *Blackjack* – and he concedes that coming up with original games wasn't part of the plan. 'The job was to get games out, fast.'

This occasionally led to playing fast and loose with rights. 'Even *Radar Rat Race* was originally an arcade game called *Rally-X*,' says Harris. 'But we didn't actually have the rights to market *Rally-X* in the UK – they only had the rights in Japan, where they created the port. So I think Andy Finkel [another of the VIC Commandos] himself adapted the graphics and made it cats and mice and cheese instead of little racers and oil slicks.'

This quick and dirty approach to creating games is what allowed Commodore to launch in the US with a bunch of cartridges, but Tomczyk was determined that the VIC-2o buyers had even more options. 'I started recruiting software developers who already had other programs,' he said. 'One guy came in and I persuaded him to change his to a colour version that became *Jupiter Lander*. We did a road race game. And then I called Scott Adams who invented the *Adventure* games.'

For those unfortunate people who aren't familiar with this landmark series, the games were text-based and put you at the centre of the action – searching for lost artefacts in *Adventureland*, waking up Count Cristo in *Voodoo Castle*, and travelling distant worlds in *Strange Odyssey*. The VIC-20 would launch with seven *Adventure* games, all ported over (they worked on the Apple II and TRS-80, but not the PET) at great effort by Andy Finkel and Adams. Adams would be rewarded in time with over $100,000 in royalty cheques.

There would be no such windfall for Finkel, but Harris describes his colleague's work on the *Adventure* games series as a 'miracle': 'That challenge was that those games were written as 24K of assembly language, and the cartridges only held 16K. So Andy figured out somehow how to shoehorn them into a cartridge. It was a small miracle.'

With so many positives, and a low $299.95 price, the VIC-20 had many ingredients for a successful launch. It even received the blessing of esteemed computer magazine Byte, which described it as 'unexcelled as a low-cost, consumer-oriented computer. Even with some of its limitations… it makes an impressive showing against more expensive microcomputers like the Apple II, the Radio Shack TRS-80, and the Atari 800.' [4]

The question, then, is why it didn't fly off the shelves when it was eventually launched in early 1981? Kit Spencer, who Jack Tramiel dragged across the ocean to help turn around the VIC-20's fortunes, soon realised that Commodore was attempting to use its existing sales channels for PET computers. 'They were trying to sell the VIC-20 through business computer dealers, and it wasn't a business computer. We decided to attack the game market through consumer channels. And some of the strategy was, "Why buy a video game when you can learn computing too?" Which is quite a powerful argument to parents.'

He adds: 'In a very short time we set up a whole consumer distribution sales network, we redid all the packaging, we just relaunched. We had William Shatner [Captain Kirk from Star Trek] as spokesman, he was obviously a well-known personality, very good for that sort of thing. And relaunched it, you know, to the home market. And it took off. And it was a price point that could be right for consumers.'

While the VIC-20 would go on to be a worldwide hit, this friendly computer would never fly off the shelves in the UK. There are no concrete records of sales, but estimates suggest it sold in the tens of thousands (which is still enough to mean it was many people's first computer). The problem for Commodore UK was that it had tougher competition in Britain thanks to existing gaming-friendly machines: the ZX81 came out in the same year, but the Sinclair ZX Spectrum would prove to be the monster success.

It didn't help the VIC-20's fortunes that the Commodore 64 would follow so quickly on its heels. With just 5kB of built-in memory, and less than 4kB available to users, the VIC simply didn't have the versatility of other computers released at the same time. While it fulfilled Michael Tomczyk's aim of being easy to use, it was a baby computer – anyone who had ambitions to do more than play games quickly moved on to more grown-up affairs.

All that said, the VIC-20 – helped along by its friendly programmers reference guide, complete with cartoon characters drawn by Tomczyk – would give many people their first taste of programming. 'Between Neil [Harris] and myself we put in some really stunning one-sentence programs,' says Tomczyk, who would later write similar examples for the Commodore magazine. 'When people saw how easy they could program and do things, they just went crazy about it. They spent time programming as a hobby.'

So what is the VIC-20's legacy? Spencer describes it as the computer that gave Commodore its 'transition to the home market, to the mass market', a view echoed by Neil Harris when he describes the VIC-20 as 'the icebreaker'. The VIC-20 on its own didn't set many people off on a career in computing, but it paved the way for the Commodore 64.

There's a compelling argument that the VIC-20 also bought the US computer manufacturers an extra year from Japanese rivals, because the VIC-20 made them stop development. Tomczyk likens it to meeting a bear in the woods. 'What do you do when a bear chases you in the woods? You take off your knapsack, you throw it down at the bear's feet, and you run like hell. And the bear stops to examine the knapsack.'

As described in that pivotal meeting in England, Tramiel knew that the Japanese were coming, and Tomczyk saw evidence of this while in Japan in the form of a 32kB colour computer made by NEC. 'Applied to the Japanese, the bear in the woods strategy works like this,' says Tomczyk. 'The Japanese are coming after us with a 32kB colour computer. So we took the VIC-20 and we introduced it in Japan first. All the Japanese skid into a screeching halt. They went back into a 12-month planning and diagnosis cycle, because they would make sure all their Is were dotted and Ts crossed before they did anything in the US market.'

Tomczyk, with his focus firmly on American buyers, believes the VIC-20 had a wider impact too. 'The VIC-20 actually jump-started the home computer revolution. At the time, in 1980, everybody was wondering, where's the home computer revolution? There were three computers that were in the market. One was the Apple family. The second one was the Radio Shack TRS-80. And the third one was the Commodore family. But there really wasn't a home market developing yet.'

The VIC-20, in essence, gave the USA what the UK already had via Clive Sinclair's ZX80 and ZX81. 'It's my contention that this was the first true home computer, it was certainly the first full-featured home computer,' says Tomczyk. 'The home market was the critical market for diffusion and adoption of computers in society. All of this came from Jack Tramiel's philosophy, which he stated many times in speeches and interviews and at the company: "I want to make computers for the masses, not the classes."' And the VIC-20 was certainly that.

Sources

Interviews with Neil Harris, Kit Spencer, Leonard Tramiel, and Michael Tomczyk.

1. Brian Bagnall, *Commodore: A Company On The Edge*, Kindle edition, location 4549

2. As above, location 3849

3. As above, location 5030

4. Gregg Williams, *The Commodore VIC 20 Microcomputer*, Byte Magazine, May 1981, Volume 6, page 46
   archive.org/details/byte-magazine-1981-05/page/n65

IBM Personal Computer (5150)

The computer that legitimised an industry

'No one would have believed in the last years of the nineteenth century that this world was being watched keenly and closely by intelligences greater than man's and yet as mortal as his own; that as men busied themselves about their various concerns they were scrutinised and studied, perhaps almost as narrowly as a man with a microscope might scrutinise the transient creatures that swarm and multiply in a drop of water.'

So begins *The War Of The Worlds* by H.G. Wells, and while some IBM executives of the late 1970s may have taken offence at this parallel – the world being watched was the explosive growth of personal computers, the eyes behind the microscope belonging to IBM – there's a ring of truth to this parallel. The main difference being that the template of the PC, as established by IBM, would outlast all of the 'transient creatures' other than Apple.

There was one other crucial difference: those creatures knew they were being watched. Everyone in the nascent microcomputer industry knew it was a matter of time before IBM would make the leap from building mainframes and minicomputers to personal computers. The only question was when.

One popular story goes that the IBM Personal Computer was kicked into action in mid-1980 when Atari sent a letter to IBM's then chairman, Frank Cary, suggesting that it could make IBM's personal computers. Rather than fling the invitation into the bin, so the stories go, Cary passed it on to Bill Lowe. Now dubbed 'The father of the IBM PC', at that time Lowe was IBM's Director of Entry Systems.

Contemporary accounts suggest this is, at best, a blurring of facts. According to Ray Kassar, then CEO of Atari, the potential partnership was instigated by Bill Lowe. 'We had two meetings actually, one in my office and another at my apartment in San Francisco with IBM,' said Kassar [1]. But the discussions never got far, most likely due to Atari's proprietary design and the fact its computers could only output 40 columns.

In truth, Lowe didn't need a memo from Atari to tell him that IBM should be building a new computer; it was something he had been convinced of for years. Why had IBM resisted? As Lowe would reflect in 2007, IBM in the late 1970s was in defence mode, 'fighting the Justice Department in the US and fighting legal battles overseas' [2] to protect its hardware and software designs and make sure no rival could service its products.

But Lowe wasn't done yet. In the 1996 documentary *Triumph of the Nerds*, he recalled his subsequent conversation with the IBM chairman. 'He kind of said well, what should we do, and I said, we think we know what we would like to do if we were going to proceed with our own product. And he said no. At IBM it would take four years and 300 people to do anything, it's just a fact of life, and I said no sir, we can provide you a product in a year. And he abruptly ended the meeting and said, you're on, Lowe, come back in two weeks and tell me what you need.' [3]

What Lowe needed, it transpired, was a team of twelve young, dedicated engineers who would work flat out for the next year. 'We were selected to go work on a top-secret project,' said Patty McHugh [4], a senior associate engineer on the team who designed the motherboard. 'Our mission was to get a product into the market in a year using off-the-shelf components.'

This was a radical departure for IBM. 'The key decisions were to go with an open architecture, non-IBM technology, non-IBM software, non-IBM sales, and non-IBM service,' said Lowe, 'and we probably spent a full half of the presentation carrying the corporate management committee into this concept because this was a new concept for IBM at the time.'

In particular, the only proprietary chip on the motherboard contained the BIOS (basic input/output system). While that gave IBM some protection against copycats, it was flimsy. Even by 1980 it was well established that other companies could legitimately reverse-engineer a BIOS – crack that, and anyone in the world could build a computer that was 100% compatible with any software that ran on the IBM PC. A platform was born.

Famously, the other key decision was to use 'non-IBM software'. In August 1980, the month in which the IBM board officially signed off on the project, it was unclear who would be providing this software, but Microsoft was already the frontrunner. Jack Sams was the engineer in charge of software development for the IBM prototype and had plenty of experience working in BASIC: he had spent months wrestling with the language in an attempt to get it working on a minicomputer (the IBM System/23 Datamaster), delaying the project by a year in the process, and had no desire to repeat the same mistakes.

By all reports, Sams liked the cut of young Bill Gates's jib. And it's worth emphasising the 'young'. While Gates was 24 by this time, he still had the physique and face of an adolescent, leading Sams to initially assume that Gates was the office boy when they met for the first time. But by the end of their second meeting, he was convinced by the young man's brains and professional manner; Microsoft was a company that IBM could do business with.

To Sams's disappointment, though, Microsoft couldn't provide the CP/M operating system that a business computer would surely need. For this, Gates told him, he would need to meet with Gary Kildall of Digital Research. Gates and Kildall had been working together for some time – they even discussed merging their two companies – so Gates had no hesitation in picking up the phone to Kildall and arranging a meeting on IBM's behalf. Days later, Sams, along with a couple of other IBM executives, flew up to meet Kildall in Pacific Court, California.

Here we move from recorded fact to disputed speculation, with Sams's meeting – or non-meeting – with Kildall being the stuff of Silicon Valley legend. Most famously, Bill Gates allegedly described it as the day 'Gary went flying', leading to the apocryphal idea that Kildall spent his day piloting his private plane for pleasure rather than meet with IBM.

There are some things we know for sure. Gary Kildall wasn't there at the start of the meeting, with his wife Dorothy initially greeting the IBM contingent; this makes sense as she handled the company's business dealings. When IBM handed over a non-disclosure agreement that one Digital Research would later describe, euphemistically, as 'unidirectional', she called in the company lawyer.

While Sams initially stated that he never met Gary Kildall, it now seems certain that the Digital Research founder returned from his business trip to Oakland (having flown there; that much appears true) and, after some deliberation, signed the agreement. Things didn't go any more smoothly from this point on, though, with IBM's stance being that it wanted to buy the rights to CP/M outright for $250,000. Kildall said no; he wanted to keep to the $10 per licence agreement he had elsewhere.

Whatever the truth of that day, we know that Sams was not impressed by Digital Research. He wanted to deal with Microsoft, and the businessman in Gates sniffed an opportunity greater than his loyalty and friendship to Kildall. Although 'friendship' may be overstating it: Kildall would write in an unpublished manuscript [5]: 'Our conversations were friendly, but, for some reason, I have always felt uneasy around Bill. I always kept one hand on my wallet, and the other on my program listings.'

Those instincts were probably correct. When Gates heard about an operating system called QDOS – the Q and D standing for quick and dirty – he realised he could provide exactly what IBM wanted without paying Kildall a penny for licensing. He also realised that the company that owned the IBM PC's operating system would have immense power; how much better for that power to be in Microsoft's hands than, let's say, Seattle Computer Products, creators of QDOS.

It helped that Microsoft and Seattle Computer Products already had a business relationship. Its most important programmer, Tim Paterson, had helped Microsoft develop an add-in card for the Apple II that would run CP/M. It was Paterson who created QDOS for Intel's new 16-bit 8086 processor, which Seattle Computer Products would soon rename 86-DOS.

But there were a couple of problems. First, QDOS was heavily based on CP/M; although Paterson didn't have access to the Digital Research code, there was publicly available documentation that allowed him to effectively mimic its way of working. Second, in August 1980, QDOS remained rough and ready, although it did include some improvements on CP/M. And third, it was owned by Seattle Computer Products: Gates wanted Microsoft to own all the software rights.

Despite these hurdles, Gates and the newly hired Steve Ballmer felt enough confidence in the operating system – and their ability to acquire it – that they could act. They duly flew out to Boca Raton near Miami, home of IBM's new PC division, and pitched for a deal that would make both men billionaires.

The most important element was ownership. Microsoft would own the DOS to run the IBM personal computer (at that point called Project Acorn, with no one involved aware of the fledgling British company's existence) but license it to IBM for a one-off fee. IBM would not be able to create its own version of the DOS, with any amendments needing to go through Microsoft. And most crucially of all, this wasn't an exclusive deal: Microsoft could also license the operating system to any other companies that came along.

'The key to the structure of our deal was that IBM had no control over our licensing to other people. The lesson of the computer industry in mainframes was that over time people build compatible machines or clones or whatever term you want to use,' said Bill Gates [6]. 'And so we were hoping that a lot of other people would come along and do compatible machines.'

As history, and Microsoft's share price, reveals, not only was Bill Gates's prediction right, but IBM said yes. It helped that there was a personal connection: Bill's mother,

Mary, had served on the board of directors of charity United Way, as had incoming IBM chairman John Opel. 'Oh, is that Mary Gates's boy's company?' [7] Opel reportedly asked when he heard of Microsoft's involvement. The deal was signed in November 1980.

There was still much to do. First, Microsoft had to acquire QDOS. With Seattle Computer Products struggling financially, the $50,000 Microsoft offered was clearly too tempting to resist (needless to say, Paul Allen, who negotiated the deal, mentioned nothing of IBM). QDOS creator Paterson would soon join Microsoft to further develop the code, which needed urgent attention: according to Gates, even on the date he signed the agreement it was apparent that they were three months behind schedule. For the next nine months, he would drive his team of around 40 programmers to the edge with countless all-night sessions to meet IBM's deadlines.

Meanwhile, in Boca Rotan, IBM's team of twelve engineers were putting in similar shifts under the leadership of Don Estridge, who had taken over the project after Lowe had been promoted. The team's existence, well away from the IBM mothership, allowed them to operate in a very non-IBM-like way: there was no big corporation clock-on, clock-off mentality here, with engineer Mark Dean describing the team as a 'tight-knit family'. 'We would celebrate together and eat together,' he said in the IBM Centennial Film, *They Were There* [8]. 'I don't think that any of us slept together but we would do just about everything else together. We would be at work late. We trusted each other.'

The hard work paid off, with Patty McHugh stating that the team shipped the first prototype to Microsoft by Thanksgiving (27 November in 1980). With Estridge and Gates in constant correspondence, using an early form of email as well as meeting up face to face, the fast pace of development continued.

Nor did the rush mean poor quality. If anything, the IBM computer was the most thought-through microcomputer design yet seen, with an attention to detail and reliability that lesser companies simply couldn't match. For instance, this was the first microcomputer to run a series of hardware tests (POST) during its boot-up sequence; it meant it took longer to start, but better for a slight delay at the start of the day than a hardware error to cause a crash during a crucial calculation.

This was also the first microcomputer to include a parity bit in the memory, which meant the hardware could detect corrupted memory before it caused potentially fatal errors in processing. According to Estridge, IBM even paid attention to the levels of contrast between the screen and the monitor's bezel to reduce eye fatigue [9]. The end result was a solid computer that businesses could trust.

Was it exciting? To those in the know, yes. The day after IBM announced what it called the IBM Personal Computer 5150 and everyone else called the IBM PC, The New York Times quoted Christopher Morgan, editor-in-chief of Byte magazine, as saying, 'It's one of the most important announcements we've seen in the industry.' [10] Michael McConnell, executive vice president of retailer ComputerLand, made this prescient point in the same article: 'People will now know that personal computers are not a fad or a flash in the pan.'

Apple, which shipped almost 80,000 computers in the US alone in 1980 [11], was a little more sniffy. In reaction to the IBM Personal Computer, it took out a full-page ad in the Wall Street Journal with the headline, 'Welcome, IBM. Seriously.'

Tandy, which dominated the US microcomputer market in 1980 with sales of almost 300,000 TRS-80 computers, appeared just as relaxed. 'I'm relieved that whatever they were going to do, they finally did it,' said the company's chief of financial planning, Garland Asher, in The New York Times article [12]. You can almost hear the sneer in his voice as he continued: 'I'm certainly relieved at the pricing. They haven't introduced anything that's going to rewrite the ground rules.'

The IBM Personal Computer cost a similar amount to the Apple III. At launch, you could buy an IBM PC 5150 with 16kB of RAM, keyboard, and monochrome monitor for $1,565. That price included BASIC, but you were expected to buy PC-DOS – the name of IBM's licensed version of Microsoft's DOS – for $40.

After Kildall threatened to sue IBM due to the similarity of PC-DOS to CP/M, the companies agreed that users could buy the Personal Computer with a version of CP/M (called CP/M-86). This arrived on the market six months after the computer's release, but it was doomed from the start. For one, it cost $240, six times the price of PC-DOS. More crucially still, by this time there was a flourishing market of software written for PC-DOS. Even at launch, you could buy popular spreadsheet VisiCalc, EasyWriter, and a number of accounting packages. For a bit of fun, you could play *Microsoft Adventure* too.

Incredibly for IBM, a company famous for taking years to create new products, the first IBM Personal Computers started shipping in October 1981. That's 14 months after it received the green light from the board. Equally unusually, it chose to sell the computers through retail stores, including ComputerLand and Sears. IBM even created a series of ads aimed at consumers – the first time in its 70-year life that IBM had communicated to anyone other than businesses – featuring Charlie Chaplin's The Little Tramp character.

While sales were initially modest compared to the likes of Apple and Tandy – it had shipped 13,000 units by the end of 1981 [13] – this was due to production limits rather than a lack of demand. Byte reported that 40,000 were ordered on the day the 5150 was announced[14], and IBM spent the next two years desperately trying to catch up with demand. According to market research firm Dataquest, it sold 156,000 computers in 1982 [15]. It's easy to argue that this success brought credibility to computers for the first time; a credibility enhanced yet further when Time magazine put The Computer on the cover of its traditional 'Man of the Year' issue in January 1983.

Veteran Byte and PCW journalist Dick Pountain was one of the first in the UK to buy an IBM Personal Computer. 'I knew it was going to be it,' he says. 'At this point I'd had an Apple IIe at home, on loan, and didn't like it at all. I didn't like any of the word processors on the Apple and I didn't like the screen. When I saw that MS-DOS was more or less CP/M, and I knew what the IBM name was going to do for the software base of it, I just thought, that's the way to go. And it was.'

IBM's honeymoon period would continue for several years to come, but it's notable that within a year Compaq released a computer that was 100% compatible with the IBM Personal Computer. That meant that any software that ran on the IBM would run on Compaq's machine. And, shortly afterwards, those of Dell, Gateway, HP, and Packard Bell. As luck would have it, a company called Microsoft, run by a beaming man-child, was more than happy to sell you the software to make it run.

Why the IBM computer was the beginning of the end

For Guardian journalist Jack Schofield, the end was written in plain letters once IBM created the Personal Computer. It highlighted a flaw in the business plan of every single computer maker of the 1980s: to keep going as a company, you needed to have hit after hit after hit.

'Commodore had a hit with the VIC-20, and then a bigger hit with a Commodore 64,' he said in late 2019. 'And after that it was bust. So I predicted that all of these companies would go bust because they couldn't hit a winner every single time. They might get one, they might get two, they might even get three in a row. But ultimately, they were all doomed because they couldn't get ten in a row.'

What's more, companies didn't help themselves by their lack of backwards compatibility. 'The logical inference at the time was that the IBM PC was going to sweep the world, because you didn't have new blockbuster machines, you just upgrade the old ones. And in fact, out of Hong Kong at the time, there was a continuous stream of improvements via plug-in cards and add-ons that gradually got incorporated into the build, as it were. And this strategy of continuous improvement meant you could run the same software, because, you know, companies were willing to support software for a successful machine.'

The IBM PC was also a sign of an industry growing up. 'When we started out, there was kind of a novelty in owning a computer,' said Schofield, 'and you were expected to write your own little programs. But once games became established, the computer became just a vehicle for running software. So if you didn't have software, you didn't have a computer that was worth anything. And you didn't have software unless the software houses believed in the future of your machine.'

This was particularly bad news for the British computer manufacturers if they insisted on creating proprietary systems. 'All of the British machines seemed to me to be doomed because they had no future development prospects.'

Fortunately for the likes of Amstrad, Acorn, and Sinclair, though, they still had a few years of life left in them – as we will see in the next few chapters.

What came next

IBM Personal Computer XT (5160)

Release 1983   **Price** £4,995

Pity the fool who bought an IBM Personal Computer 5150 in mid-1983: by the end of that year, you could buy a PC with a 10MB hard drive built-in. Journalist Dick Pountain remembers buying a 5MB external hard drive for £1,500 all on its own for his Model 5150 PC.

IBM PCjr

Release 1984   **Price** £1,269

On paper, the PCjr was an excellent idea. Attract the home buyer with a colour screen and essentially the same specifications as the first IBM PC. Except that it wasn't fully compatible, so you couldn't guarantee all software would work, and everyone hated the original chiclet keyboard.

IBM PC Convertible

Release 1986   **Price** £2,500

IBM had made two different luggable computers – the original 5100 way back in 1975 and the ill-fated 5155 in 1984 – but the PC Convertible was its first notebook computer (and also the first IBM PC to feature a 3.5-inch floppy drive). It could run on batteries, but was hampered by a monochrome 640×200 panel display.

IBM Personal Computer AT

Release 1984   **Price** £6,000

A 20MB or 40MB hard disk, along with Intel's 6MHz 80286 processor, were the stand-out features of the AT, which was designed to sit at the top of IBM's range. PC-DOS 3 was something of a disappointment, though, and by this point clones were appearing in their droves.

IBM Personal System/2

Release 1987   **Price** from $2,295 (Model 30)

In a last-ditch attempt to take back control of the PC market it had created, IBM created a proprietary architecture called MCA (micro channel architecture). Which meant that, on the high-end models that included it, IBM created a sub-category of PCs that weren't actually compatible with PCs. It also announced its own OS/2 operating system, which it's safe to say did not take off as IBM hoped.

Sources

Interviews with Dick Pountain and Jack Schofield. Total number of IBM Personal Computers sold based on IDC figures published in InfoWorld, 30 March 1987, page 1.

1. *Atari Inc. Business is Fun*, Kindle edition, location 4975
2. William Lowe, Commodore 64 – 25th Anniversary Celebration, 10 December 2007,
    youtu.be/NBvbsPNBlyk?t=2498
3. *Triumph of the Nerds*, John Gau Productions, first broadcast on Channel 4, 14 April 1996
4. IBM Centennial Film: *They Were There*, interview starts at 18:55
    ibm.com/ibm/history/ibm100/us/en/films/theywerethere.html
5. Gary Kildall, *Computer Connections: People, Places, and Events in the Evolution of the Personal Computer Industry*, page 62
    computerhistory.org/blogs/computer-history-museum-license-agreement-for-the-kildall-manuscript
6. *Triumph of the Nerds*, see 3
7. Paul Freiberger, *Fire in the Valley: The Birth and Death of the Personal Computer*, 1999 McGraw-Hill edition, page 334
8. IBM Centennial Film: *They Were There*, interview starts at 19:28
9. Lawrence Curran and Richard Shuford, IBM's Estridge, Byte Magazine, November 1983, Volume 8, Issue 11, page 88
    archive.org/details/byte-magazine-1983-11-rescan/page/n89/mode/2up
10. Andrew Pollack, *Big I.B.M.'s Little Computer*, The New York Times, 13 August 1981, page D1
11. Jeremy Reimer, *Total Share: Personal Computer Market Share 1975-2010*, 7 December 2012
    jeremyreimer.com/rockets-item.lsp?p=137
12. Andrew Pollack, *Big I.B.M.'s Little Computer*, The New York Times, 13 August 1981, page D6
13. Paul Freiberger, *Fire in the Valley: The Birth and Death of the Personal Computer*, 1999 McGraw-Hill edition, page 348
14. Sol Libes, Bytelines, Byte magazine, volume 6 number 12, December 1981, page 314
15. David Thomas, *Alan Sugar: The Amstrad Story*, Century 1990, page 221

The screen displays:

```
BBC Computer
Acorn DFS
BASIC
>LIST
>
>
```

BBC Micro

The computer that taught
the UK to code

'They are marvellous at making programmes and so on, but by God they should not be making computers, any more than they should be making BBC cars or BBC toothpaste.' [1] So said Clive Sinclair in a typically combative tirade in 1982, shortly after the BBC Micro started shipping. In truth, the Micro was a brilliant, bold decision by the BBC in its pomp – and one that would have an immeasurable impact on the nation's schoolchildren.

It could have all been so different. The BBC, and the UK government, had largely ignored the rise of computers during the 1970s. This changed when the BBC broadcast a Horizon programme called *Now The Chips Are Down* in March 1978, which ended with a famous scene as the screen faded to black. 'What is shocking is that the government has been totally unaware of the effects this [microprocessor] technology is going to create,' said the narrator. 'The silence is terrifying.'

According to David Allen, who would become the producer of the BBC Computer Literacy Project that commissioned the BBC Micro, the documentary and its damning verdict 'catalysed a lot of things' [2]. In particular, the Callaghan government asked the BBC if there was anything it could do to raise awareness. And that led to the then BBC Controller for Educational Broadcasting, Sheila Innes, to famously ask Allen if 'there was anything in it?'

Kick-started with £10,000 from the government's influential Manpower Services Commission, Allen initially created a three-programme series to cover the rise of microcomputers. Called *The Silicon Factor*, and broadcast in the spring of 1980, it provided an excellent taste of what was already happening in the technology industry – and the impact it could have on our world.

According to a BBC report published in 1983 [3], this trio of shows also changed the tone of a watching nation's questions from 'How will this affect my job, or my company, or my industry?' to more practical questions such as, 'What is a microcomputer?', 'What is a computer language?' and, most importantly, 'How can I control a computer?'

'These questions became the basis of what was to become the BBC Computer Literacy Project,' stated that same BBC report. 'In November 1979, it was agreed that we should make a ten-part television series for adults… to be broadcast from October 1981.' Crucially, the planned series would have a practical, hands-on element.

But there was a problem. The BBC's advisors, most notably John Coll, who was the Chairman of the Micro Users in Secondary Education (MUSE), explained that they needed a standardised version of BASIC so that people could type example code and be certain it worked. He convinced the BBC that it should adopt a version called ABC, developed by MUSE, which stood for Adopted BASIC for Computers. 'It was very structured and academically very acceptable,' said Allen in 2018 [4], 'but no machines actually implemented it.'

The BBC invited a number of British computer manufacturers to London for a meeting to see if they would adopt ABC. 'They said, if the DTI [Department of Trade and Industry] will pay for the ROMs or whatever it is that we do, we will do it,' said Allen. 'Failing that, we won't.' And so ended that idea.

Fortunately, or so it seemed at the time, the government already owned a computer company. When the much-maligned National Enterprise Board (NEB) had severed ties with Clive Sinclair in the late 1970s, the DTI had taken ownership of the work his company had already done on a Zilog Z80-based computer codenamed NewBrain. Perfect: the BBC would adapt this computer to fit its needs, and there could be no accusation that government money (via the BBC) was going into the hands of a private company.

There was only one bug in the BBC's immaculately planned code: after six months of little progress, it became obvious that Newbury Laboratories could not produce the NewBrain on time. 'We approached Newbury, and they said yes, we can put ABC [on it] and it can do all the things that we want,' remembered Allen. 'And they went away and they tried to produce a machine and effectively they failed. It was rather sad. They got something that could do some things but when you got it to do other things it just wouldn't work, it kept crashing and so on and so forth.'

In an act of 'desperation', John Coll and David Allen drew up a specification of what they wanted their ideal machine to do over Christmas 1980. The key elements? 'Colour, graphics, sound, a fully positive keyboard, the ability to run a proper monitor, but also run on a domestic television,' said Allen. It also needed to be used in a TV studio so that its output could be part of the broadcast.

Now the BBC's management had to make a big decision: press ahead with the Computer Literary Project without a computer running its version of BASIC, or create one itself as specified by Coll and Allen. To an extent, this was a natural extension of what the BBC had always done to support its further education output: the programmes were accompanied by publications to help people learn. The plan was always to do the same with the computer literacy programming. Now, there would merely be a BBC computer to accompany it.

To its credit, and with due acknowledgement of executive producer John Radcliffe's persuasive talents, the BBC senior management team backed the idea of a BBC Micro. So just one small thing to decide: who would make it?

Radcliffe approached the DTI for its recommendations of existing computer British manufacturers, with a proven track record, who could do the job. It produced a seven-strong shortlist comprising of Acorn, Nascom, Newbury, Sinclair, Research Machines, Tangerine, and Transam.

All but one company decided to throw their hat into the ring and Mike Fischer, co-founder of Research Machines, told us why: 'We were approached by the BBC to bid for their concept of a BBC computer, but we felt the timescale and price they had in mind weren't achievable.'

Christopher Curry, co-founder of Acorn, had no such concerns. In fact, the moment he had heard about the BBC microcomputer project – well before Christmas 1980 – he had been putting pressure on the BBC to open up the tendering process. But not before he phoned up his former boss, Clive Sinclair (see the ZX80/ZX81 story, page 48). 'I rang him up and said, "What are you up to at the bloody BBC?"'

Curry was passionate about this because, in his words, 'if you get the BBC working with you, you've got the biggest advertising organisation we could never afford for nothing.' Curry still remembers Sinclair's response. 'He said, "I'm not doing anything. Trust me, I'm not." And I didn't believe him. Then we found that the likely ones were the people in Newbury.'

That conversation happened much earlier in 1980. While Newbury were still in the official running come January 1981, Curry's hopes for an open process were now realised and Acorn was determined to win the contract. 'They came to 4A Market Hill,' says Hauser. 'They told us the specification that they wanted, with everything including the kitchen sink in it. And as luck would have it, Steve [Furber] has a design in his drawer, which was very similar to what they wanted.'

Hauser is referring to the Acorn Proton, the successor to its Atom (see 'The mighty Atom'), but Sophie Wilson points out that it wasn't even at the design stage by January 1981. 'We had nothing,' she says. 'We agreed the "Proton" would be a professional computer, but couldn't agree on the CPU. In December 1980, I came up with the idea of it being a multiprocessor machine sold in sections, a 6502-based I/O processor first, which met with approval. But we didn't have a design.'

At this point, we should stop to reflect on the fact that Acorn employed two people who would go on to be recognised as giants of the technology industry. Together, Steve Furber and Sophie Wilson would not only be the main force behind the BBC Micro but also create the ARM processor that powers all modern smartphones. As Oscar Wilde would say, to have one industry-shaping genius in your company may be regarded as good fortune; to have two looks like selfishness.

Along with many other talented people who don't get the recognition they deserve – Paul Bond, Jon Thackray, David Seal, and Kim-Spence Jones to name just four – Acorn had one more trick up its sleeve: a couple of canny bosses. 'I rang Sophie and said, you know, Sophie, is there any chance we could do this by Friday?' says Hauser. 'And she said absolutely no chance. Forget it. So I rang Stephen and I said, "Steve, I've just been talking to Sophie, you know, and she says, if we really try hard, we might be able to get it done by Friday." "Absolutely out of the question," says Steve, "but if Sophie is in then I'm in."'

Hauser then rang Sophie back and fed her the same line. Naturally, his artifice didn't last for long. 'That evening, we didn't have a clue that we'd been conned,' says Wilson. 'On Monday morning, we found out very quickly. But we'd already agreed by then.' Wilson insists there were no hard feelings: 'That relationship between me, Steve and Hermann was very close – by then we'd been working together for a couple of years.'

The next four days were arguably the most important in the history of modern British computing. They were certainly among the most frenetic. While Furber had a clear idea of how he wanted the Proton to work – it was based on a design he'd created at home the previous year – there were 'significant enhancements such as doubling the memory rate' to contend with. Plus, at this point the team only had a block sketch of the machine's integrated circuits: they needed a detailed circuit diagram. 'That was basically Monday and Tuesday,' says Wilson.

At the same time, Acorn had to source the cutting-edge components. 'The BBC Micro has high-speed memory shared between the video system and the processor,' says Wilson, 'so we needed the highest-speed DRAM currently in existence. We'd seen them in a Hitachi data book but there were none in the country. I believe Hermann got in touch with Hitachi head office in the UK and convinced them, and their head salesman in the UK hand-carried some samples to us.'

They also needed to source a fast, 2MHz 6502 processor to support the 4MHz memory, which again meant calling in a favour from a rep. But the biggest challenge of all was to create the prototype. With 3,000 connections, the only option was to wire-wrap it. Enter Cambridge University's Ram Banerjee – 'the fastest gun in the west – he could wire wrap the prototype board faster than anybody,' says Hauser.

This took most of Wednesday, and even with Banerjee's expertise there were areas where the wire wrap didn't make contact correctly. The team spent Wednesday night and most of Thursday finding then correcting the mistakes, with Hauser providing a constant supply of food and tea.

'By Thursday evening, we had a prototype that looked as though things were OK,' says Wilson. They hooked it up to an Acorn System 3 computer with an in-circuit emulator (ICE) card inside, allowing them to emulate all the actions of the processor. 'Through that we gradually proved that everything in the machine really was connected properly. But the thing was resolutely refusing to work.'

By now it's 2am on the morning of Friday 20 February, with the BBC representatives due to arrive at 10am. And Wilson still needs to write the code that will allow the prototype to actually function; wisely, she heads home to get some rest, leaving the rest of the technical team to diagnose the fault.

'I changed my job from tea lady to explaining to them that the electronics would only work if they set up a link with a clock,' says Hauser (the connection to the Acorn System 3 via the ICE link was producing a 'clock skew' in the prototype). 'Let's just forget about the ICE link. Let's blow the program into the ROM* and fire it up all by itself. And this was the last thing that we could try and they said this is hopeless, this is stupid. But since I was the boss, they did it anyway. And it worked.'

At 8am, Wilson walks back into the office and is assured that the prototype is working. 'So I ported Acorn System BASIC and enough of [the operating system] to the board. And the BBC arrived.' Cue a stalling game, as the Acorn directors

---

* To be absolutely correct, there was no program in the ROM. 'I hadn't written it yet,' says Sophie Wilson. 'They just needed to get the processor and memory system running properly, testing it with a scope.'

delayed the BBC entourage downstairs while Wilson attempted, typing blind, to set up the monitor's controller. 'By the time they got upstairs, there was BASIC running on the machine prototype and the Mode 0 display was displayed,' recalls Wilson. She even had time to set a demo program running, showing a random line walking across the screen.

It was nothing short of a triumph.

In stark contrast, the BBC's delegation was not impressed by its visits to other manufacturers. Allen described most of Acorn's rivals as 'box movers' in 2016 [5], but the biggest disappointment was Sinclair. 'The ZX80 and the ZX81 were on the streets by then but they weren't the sort of machine we wanted – they were effectively too flimsy, too lightweight, and with not enough expandability.'

The ever-secretive Clive Sinclair refused to show them the under-development Spectrum, but Allen wasn't impressed by what he did see. 'When we said we wanted a fully positive keyboard, by which we meant solid keys, he waved the Spectrum flexible keyboard at us, saying that's a fully positive keyboard, which we didn't really think was the case.'

The contrast between Acorn, which had essentially whipped a prototype out of thin air within a week to meet the BBC's requirements, and Sinclair is stark. Thanks to Hauser's quick thinking, and industrial designer Alan Boothroyd's equally quick work, they had even been able to show a rough model of what the case would look like.

Curry also points to Acorn's heritage, heading all the way back to December 1978 when he and Hauser set up their first company, Cambridge Processor Unit Ltd. It soon won a contract to build a microprocessor-controlled one-armed-bandit. 'Steve Furber was convinced that the way to make it was using a twin processor design that allowed the massive amount of I/O required by a fruit machine to be handled by the second processor and create an interface bus called the Tube.'

This had been factored into the design of Proton. 'When I first explained to Richard Kitson and John Radcliffe that the Proton could run on both Z80 and 6502, and include the structured BASIC that they wanted, NewBrain almost disappeared from consideration.'

Acorn also had a ready-made way to connect computers. 'Once ensconced in Market Hill we depended on a serial comms line for moving code between developers,' says Curry, referring to its Econet local area network technology. 'It was included in the Atom with special remote screen and keyboard instructions to make a classroom teaching aid for programming with up to 20 students. This alone was a clincher for the BBC.'

It was also obvious that Acorn and the BBC Engineering team shared a similar desire to build an expandable, high-quality system. What's more, they had the skills and willingness to work collaboratively with the BBC's Engineering team – skills and willingness that would be pushed comprehensively over the next few months of development.

It wasn't long before Christopher Curry heard through the grapevine that the BBC would be awarding Acorn the contract.

While Chris Turner, who was then chief engineer at Acorn, insists that no one was in overall charge of the BBC Micro project – 'it was just a team' – he is willing to be described as 'custodian of the circuit diagram'. At this point, Steve Furber wasn't even on Acorn's staff; he was employed by Emmanuel College as Rolls-Royce research fellow.

The team had managed to create a wire-wrapped prototype of the BBC Micro within a week, but there was a huge amount of work to be done. 'Obviously, the prototype was quite limited in what it had,' says Wilson. 'No PAL encoder, no tape interface.' Once you factor in these additions, and what was already included, there were over a hundred chips in the full design.

Turner soon realised that there wasn't room to fit all the circuits onto the board within the desired case design, and stripping back the specification wasn't an option. With the help of Acorn director Andy Hopper and Furber, who Turner describes as 'popping in at lunchtimes and evenings, but thinking a lot in the interim', Turner commissioned Ferranti to create two gate array chips that would sweep up the video display processing and various logic functions. (Ferranti was the UK's pioneer in making gate arrays, which it called Uncommitted Logic Arrays or ULAs.)

Turner is keen to emphasise that the architecture is Furber's, not his, but believes that together they created something 'magical'. Perhaps the most inventive step was how it shared memory. The conventional wisdom was to have two separate memory systems: one for the graphics, one for the display. Furber's idea was to share the fast, expensive DRAM so that the processor could use it to run code and write the display memory while the graphics controller could alternately access it to refresh the display output.

Even today, there are elements of the Micro's design that Furber describes as 'scary'. In what sense? 'In the sense that we didn't fully understand why something we did to make it work actually made it work,' he explains. 'The most famous one was when we got the first PCB [printed circuit board] prototype. It was not working reliably and I found if I stuck my finger on the underside of the board in the right place, it would start working reliably. And, you know, we didn't have enough fingers to put one in every product.'

They solved this conundrum by 'putting the equivalent of "Steve's finger" in the form of a resistor pack across the 6502 data bus,' says Furber. 'This was just a set of fairly weak pull-up resistors, but that turned it from unreliable to reliable for reasons that nobody was quite sure of.'

There were plenty more challenges along the way. The biggest, and which ultimately delayed mass production of the BBC Micro by several weeks, was the Ferranti ULA video processor. Andy Hopper takes up the story. 'We had the graphics chips prototyped by Ferranti in some ceramic packages [and] it all worked fine. We ordered several thousand, which came in plastic packaging. And horror of horrors, the thermal properties on the [cheaper] plastic packages meant that the chip heated up more.' As the ULA got warmer, the pixels disappeared. 'So you switched on your computer and it's all working, and then five minutes later, there's no graphics. You can't see anything.'

In desperation, Hopper even took boards home and put them in his Aga, using physical probes to try to work out what was going on. Acorn achieved some success by adding heatsinks and speed-selecting ULAs that had better thermal properties, but these were stopgap measures – just enough to get the first batch of Micros to customers. Acorn would need to convert the ULA design to work as CMOS logic for the problem to be properly solved. (CMOS stands for complementary metal-oxide semiconductor, but its importance here stems from its reliability at higher temperatures.)

Even then, though, Acorn hit a potential roadblock. The replacement video processor chips were being made by VLSI Technology in California, so Chris Turner flew over to test the first batch. This was in November 1981, by which point Acorn should have been shipping BBC Micros to customers. Christopher Curry had already made an awkward phone call to executive producer John Radcliffe to explain a further delay.

'You can imagine by then we are under intense pressure,' says Turner. 'We had lots of orders, we had the BBC saying, you know, to move faster. And we have a production site on standby. So we've got boards assembled waiting for these chips to go into them.' But great news: when Turner arrives at VLSI to test the new chips, everything works. 'I was exercising the machine and running various test programs that we had and getting increasingly confident. And then, much to my disappointment, I discovered that the graphics cursor was inverted in Mode 7 only.'

This was the Teletext mode, which Turner had added into Furber's design – as luck would have it, he had several years of experience in television electronics to fall back on. He took a closer look at the circuit and immediately spotted the problem. 'At that point, you're faced with respinning the chip, which is going to take another few weeks. And then I looked at the circuit some more and I thought, ah, if I cut this pin here and lift the leg of this chip and put a blue wire on the board from here to here, I can turn the graphics cursor the other way up.' After checking his thinking with Furber back in Cambridge, Turner instructed VLSI to press the button and fabricate the wafers. Within a couple of weeks, the flawed Ferranti ULAs were history and the BBC Micros could start shipping in high volume.

There is a notable corollary to all these problems. Furber was well aware that the BBC Micro was always operating close to its thermal limits – 'the characteristic of the Beeb was that if you put it in a thermal chamber and heated it up to about 35 degrees, everything fell apart,' he says – and that made him determined to avoid this problem when later designing the ARM (Acorn RISC Machine) chipset. 'If you took a prototype Archimedes circuit board and stuck it in a thermal test chamber, you could run the whole thing at above 100 degrees and it would still operate reliably.' A useful property.

If you were an early customer of the BBC Micro, there was one other ever-so-slight problem to be aware of: it might overheat. This had nothing to do with the video chip and everything to do with the BBC's initial insistence that Acorn use linear power supplies. 'The BBC at that time did not like the more efficient switch-mode power supplies because they switch at radio frequencies and radio frequencies are the BBC's domain,' says Furber. 'So the first BBC Micros to ship had linear supplies… and you couldn't expect a linear supply to be more than about 50% efficient. This just generated too much heat for the space allocated in the case and so we had a number of fairly high-profile episodes of [Micros] overheating, with the possibility of setting light to things. One of which was in a hospital, I think, and it's a very bad thing to have electronics start fires in hospitals, it turns out.'

The BBC conceded this point, Acorn sourced a highly efficient switch-mode power supply, and the fire risk was a thing of the past.

Another source of creative friction between Acorn and the BBC concerned BASIC. Wilson had already started work on the next version of BASIC and had big ambitions. 'Acorn's BASIC at the time was quite radical – it was approaching a line-numberless world. It had labels as well as line numbers and I was in the process of taking line numbers out,' says Wilson. 'It had procedures and functions and things like that. So we actually had to go slightly backwards while keeping all the advanced stuff, which they decided they liked, after all.'

The BBC's input into BASIC, coordinated by Richard Russell from the BBC's Engineering department, centred around compatibility. It wanted not only a structured version of BASIC, but one that would be largely compatible with Microsoft BASIC.

'I chaired the standards committee, if you like, between Acorn and the BBC of what the commands in BBC BASIC ought to look like,' says Hauser. 'And these were not always easy meetings because [Sophie and Richard] didn't always agree on what BBC BASIC should be like, but Sophie normally prevailed because she knew this, of course, backwards and she knew what she could implement efficiently. BBC BASIC then became one of the most celebrated BASICs ever, I think, because it was a particularly nice language to write in. And it was an easy way of getting to the assembler as well.'

But we shouldn't understate the role of the BBC in the creation of the Micro. With its name in big lights, it knew the BBC Micro had to be a success, had to be reliable, and had to meet the high expectations of the public. 'The BBC did have quite a lot of input into the spec, particularly on the software side, but some on the hardware side,' says Furber. 'There were differences of opinion, which I think managed to emerge as constructive conflicts rather than just conflict.'

In particular, the BBC had ambitious hopes for what could be squeezed into an affordable machine, with what Furber describes as a 'conflict between their spec and their expectations of price'. For the most part, though, specifications won.

David Kitson, a senior manager in the BBC Designs department, has similar memories. 'One of the features of the BBC Micro was the number of connections that it had for various peripherals, which really raised it quite a long way above any competitive machine of the time,' he said in 2018 [6]. 'I don't really know how much of that was down to people like [David Allen], people like Richard [Russell], people like Hermann [Hauser], so I really can't apportion credit for that, but it was certainly an outstanding feature of that machine.'

Kitson's key contact at Acorn was Chris Turner; Kitson describes Turner as the 'unsung hero' of the BBC Micro, and not just due to his dash to California to rescue the video chip. Turner had oversight of the Micro's hardware development, production engineering, and choice of component suppliers throughout 1981. Kitson had the pleasurable task of coordinating the weekly meetings, as they came to be, to discuss progress and make key decisions. 'There were things like the design of the keyboard, the ten red keys, where the BBC logo and the owl would go on the casework,' says Turner. 'Very often, I was the one providing evidence or taking the notes and when there was agreement, making sure that was what everyone went away and did.'

Meanwhile, the BBC team back in London needed to work on pilot programmes. And that meant having a working prototype by September 1981. This, it's fair to say, was stressful for all concerned. Hot studio lights and complicated filming demands – the Micro needed to output straight to camera – do not mix well with a prototype computer prone to overheating, with roughly half the booked studio sessions going to waste as they couldn't get the Micro to work as it should.

To help, Acorn despatched Furber and Wilson to the studios so they could troubleshoot during filming, with Furber needing to try to cool the video processor down using a can of Freez-It spray. And this wasn't the only smoke-and-mirrors trick, with the BBC Micro box on the table being for display purposes only: the actual computer was hiding underneath. 'This was because it had a big heatsink on the ULA and was connected to an Acorn System range machine acting as its disk drive,' says Wilson.

Such an ambitious design ultimately led the series' broadcast date to be pushed from September 1981 to January 1982, which at least gave the BBC time to finish its supporting documentation.

In late 1981, when Acorn started taking orders for the Micro, the gap between anticipated level of demand and the real level of demand became clear. The BBC's original agreement with Acorn said the company should expect to sell

12,000 Micros over the course of twelve months. 'Steve and I thought they were extremely conservative,' says Wilson. 'We thought it would be a fantastic success and sell 50,000.'

Both estimates were wildly wrong. By Christmas 1981, Acorn had already received orders for almost 12,000 computers. Hauser estimates it sold over 100,000 machines in the first year; by the end of its life, the BBC Micro would sell around 1.5 million units.

Demand was so great that the BBC adjusted its broadcast schedule for a second time. It had planned to start the weekly broadcasts of *The Computer Programme* on BBC 1 on Sunday 10 January 1982, but it switched to BBC 2 on Monday 11 January, at the school-friendly time of 3.05pm. More mainstream times would wait until the following month.

Some of the Micro's popularity was, indirectly, thanks to the UK government. It started with an ambitious young politician named Kenneth Baker in 1979. 'Being on the backbenches and a businessman, I took an interest in computing,' he said in a Radio 4 interview in 2016 [7], 'and in the 1970s I went to Japan a lot to look at their VLSI intensive chip development that they did. I got to know what was happening in computing... I got to know a lot about all of that world. And I decided that you needed a minister to actually promote the changes that were just about to happen.'

He wrote up a ten-point manifesto for a new position of Minister of Information Technology, and presented it to Prime Minister Margaret Thatcher. Along with giving financial support to early robotic development, and the optimistic proposal of a paperless office in Whitehall, one crucial element of his plan was to put a computer into every school in the country. In 1981, by which point the UK was in a deep recession and an austerity budget in place, 'Margaret wanted something nice to say,' recalled Baker, 'so I gave her a little package of technology measures, one of which was computers in schools.'

By late 1981, this idea had developed into a full-blown scheme from the Department of Industry, where it would subsidise the purchase of approved computers by up to 50%. One of those computers was the Research Machines 380Z, the other the BBC Micro.

Another challenge: the BBC Micro had already received positive reviews in the specialist press. Practical Computing described it as 'more advanced than anything the American or Japanese can offer at this price' [8] while Personal Computer World, which had been keenly following the Micro's progress since the first rumours of its existence emerged, dedicated the front cover of its December 1981 issue to what it described as 'BBC Computer: Auntie's Micro'.

Reading that early coverage in the specialist magazines, one thing is obvious. The perception at the time was that this was essentially just another computer. Yes, it had the backing of the BBC, and that gave it instant gravitas and newsworthiness, but judged with the benefit of 40 years' hindsight it's obvious that the BBC logo was far more than a badge. For the BBC, the Micro was one cog in its ambitious Computer Literacy Project. It wasn't there to flog computers; it was there to educate a country.

The Welcome Pack was another sign of the BBC thinking like an educator, compared to the traditional 'build it and they will come' approach of computer manufacturers. This cassette tape included a series of 16 programs that demonstrated the various skills on show with the new computer, whether that was manipulating databases, creating graphics, or playing a game or two.

Again, we must give credit to the BBC for its integrated thinking. History may make it seem as if Acorn chucked a computer together and the BBC made a few TV programmes, but one of the reasons the BBC Micro has such a huge legacy is the coordinated work done to make the BBC Computer Literacy Project a success. Between 1980 and 1988, it produced 145 programmes and 166 programming exercises to accompany them.

The BBC also stretched its tentacles into the wider community. Interested adults could sign up to the 30-Hour BASIC course run at colleges throughout the country, and it was successful too: by 1983, 160,000 people had signed up to the course with a record low dropout of 2.6%. The BBC's Broadcasting Support Services provided additional advice via post and telephone, with over 300,000 enquiries made by mid-1983.

It also worked with independent developers, and Acorn, to produce software that would run on the BBC Micro. Along with the BASIC-based programming aids you would expect, there was the UltraCalc spreadsheet, a selection of games (including *Dr Who: The First Adventure*), a tax calculator produced in conjunction with *Which?*, and many more. Acornsoft, under the inspired guidance of David Johnson-Davies, would go on to publish over 20 education software titles, 15 pieces of business software, and almost 50 games – the most famous of which, of course, was *Elite*.

All this concerted effort meant the BBC and Acorn had a runaway hit on its hands. Soon after broadcasts began, Chris Turner, Sophie Wilson, and Steve Furber were invited to take part in a seminar dedicated to the BBC Micro being run by The Institution of Electrical Engineers (the IEE, now known as the IET or Institution of Engineering and Technology). But even the IEE's grand London headquarters proved too small. 'This lecture theatre held about six or seven hundred, and in the event three times the number they could legally fit in turned up. Including people who had taken coaches from Birmingham and so on,' says Turner. 'I think that was the real point at which we began to sense that this was a very big deal.'

As a result, they held two more seminars in London and took 'the show' on the road, attracting large crowds for a technical talk in cities around Britain and even Ireland. This wasn't just because of the glamour stemming from a BBC logo on the front of the PC. The Micro was an incredibly advanced machine for its time.

For instance, thanks to Andy Hopper's work, it included the Acorn Econet. While not new to the BBC Micro – it had debuted in Acorn's System 2 in 1980 and also featured on the Acorn Atom – this offered schools an easy way to link up their Micro computers and even allowed teachers to view a child's screen.

Then there was the Tube, an ingenious way to hook up a second processor. You could even add a module that would allow you to run CP/M on a Zilog Z80 chip, but Wilson dismisses this as 'one of the worst options! The 6502 processor and later the ARM second processor did better. *Elite* even had a second processor mode with more features and a higher resolution.'

All these plus points meant that the BBC allowed Acorn to increase the price of the Micro. It launched at £235 for the Model A and £335 for the Model B, but in January 1982 those prices jumped to £299 and £399 respectively. According to a BBC announcement of the time, this was due to rising costs for production, components, and testing. However, Turner thinks this was also necessary to accommodate some profit margin for all the customer-facing computer shops who were needed to stock, sell, and service the Micros.

This price rise had little effect on demand and Acorn still faced huge problems to meet supply. By June 1982, according to the first issue of *Acorn User* magazine, it had only shipped 10,000 Model A Micros and 7,000 Model Bs. Smoothing production at the two British-placed plants – part of the BBC's conditions was that the computers had to be assembled in the UK – was going to keep Chris Turner and many others in Acorn's growing workforce busy for several months to come.

There was another stipulation that would frequently annoy Christopher Curry. 'The name Acorn was never mentioned in the BBC programmes, which I found infuriating because if they mentioned the NewBrain product they would mention NewBrain, if they mentioned a Sinclair product, they mentioned Sinclair,' he says. 'But because we were the chosen ones we had to remain anonymous.'

Nevertheless, there's no doubting which British computer company most directly benefited from the BBC Micro. Acorn's turnover increased by a factor of 30 in the space of two years, and there was a meteoric lift in its profits too: in 1979, these amounted to £3,000; in the year to August 1982 to July 1983, that total hit £8.6 million.

The Micro did have its critics, with technology journalist Dick Pountain describing the Model A as a 'terrible mistake' due to its lack of memory. But more crucially, he felt it was wrong to tie the Micro to BASIC. 'Their idea was that it was going to be the educational computer, which it did become in a way, but it meant teaching all the kids BASIC, their BASIC, which was a very non-standard BASIC. And I just felt [the Micro] should have been a more open platform. They were half-heartedly following the Apple model rather than the IBM model, by being a closed, proprietary system.'

Pountain isn't alone in criticising BASIC as a learning language, and there's a reason why it isn't taught in schools any more: if you want to teach young children the principles of good programming, the visual, block-based language of Scratch is a far better starting point. Once children become more adept, or simply older, Python is now a common choice due to its easy-to-understand syntax. And let's not forget, there was a reason that Sophie Wilson was in the process of removing the line numbering from Acorn's BASIC before the BBC came along.

Nevertheless, it's as hard to overestimate the effect of the BBC Micro as it is to quantify it. For a start, unlike any of the other computers in this book, it wasn't simply a clever box: thanks to the backing of the BBC, it was supported by TV programmes, by excellent guides, by training for teachers, and by local colleges. It also had ubiquity: in the mid-1980s, it would have been hard to find a British school that didn't contain a Micro.

As a result, this was the first computer that many Britons encountered. Some used it to play games, some educational programs, others created text-based adventure games and learned the principles of programming. This inspired a generation of computer-literate children, arguably the most computer-literate children in the world at that time. Is it any coincidence that Britain remains a powerhouse of games creation and innovative tech thinking?

This is why, right at the start of this chapter, we talked about this being a brilliant and bold move by the BBC. It wasn't without negative effects – Research Machines founder Mike Fischer told us that it nearly brought his company to bankruptcy – but the BBC Micro's net contribution to the United Kingdom is seen in the ARM processor, the size of our tech industry and, perhaps most directly, the Raspberry Pi project.

Commodore and Sinclair taught a nation how exciting computers could be. The BBC Micro taught a nation how to code.

Model A or Model B?

One of the BBC's core targets was for its computer to be affordable, which is the key reason why Acorn sold two models: the Model A for £235 and the Model B for £335 (their original prices). You can almost see the company stretching to hit a sub-£200 target. But this wasn't to be: rising component and production costs meant that the BBC agreed with Acorn to increase prices to £299 and £399 in January 1982.

The key difference between the two machines was memory. The Model A came with 16kB, the Model B with 32kB (although it was possible to upgrade the Model A's RAM yourself). This extra memory meant the Model B supported more screen modes, too: while the A peaked in Mode 4, which was a generous-for-the-time 320×256 resolution, or 40 by 32 text, the B offered Mode 0 with its 640×256 resolution and 80 by 32 text.

There were also a number of functional differences that made the Model B a more suitable choice. It included a Centronics interface for printers, Acorn's proprietary Tube interface for adding a second processor, its Econet for networking other nearby Model B Micros, and a more advanced four-channel analogue-to-digital converter.

While the Model A initially sold well, it was the more flexible Model B that proved to the far greater success.

The mighty Atom

Without the Acorn Atom, there would be no BBC Micro. This was Acorn's first computer aimed at consumers rather than businesses, and while in some ways it was a simple machine compared to its successor, it was full of innovative features for the time.

'Right from the start, we decided to build in digital networking capability, which didn't really exist anywhere,' says Christopher Curry, who – along with Nick Toop – designed the Atom. This networking capability, which would become so familiar to teachers using the Micro in a classroom, was the Econet.

Sophie Wilson wrote the Atom's BASIC interpreter and this was equally innovative, as it included an assembler for the processor's assembly language – useful, because it meant you could write a program in BASIC and execute assembly code from within it. A notable feature of BBC BASIC too.

The Micro's choice of the 6502 processor also stems directly from the Atom, and the Acorn System range before it. 'All the way back to the System 1,' says Wilson, 'which, since I designed it, happened to use the microprocessor I chose for my own computer.'

Wilson was not alone. In the late 1970s, Acorn's options were either the Zilog Z80 or the 6502. The Cambridge computer community had essentially sided with the 6502. If you were building a computer in Cambridge, and wanted to tap into that amazing wealth of knowledge, it was the only logical choice.

But the Atom's most expensive part wasn't the processor but the keyboard. This was vital, Curry felt, to give the Atom a grown-up feel. He went hunting for a suitable unit, eventually finding one in Hong Kong. At $11, it was 'three, four times as much as the main processor, but I think it was a good investment and I'm glad we did it.'

Alan Boothroyd then set to work building a case around the keyboard, and with some slick photography and marketing by Curry – one of the key things he learned from former boss Clive Sinclair – the Atom was set to be a modest commercial success. With prices starting at £140 in kit form, £174.50 assembled, it was affordable too (you will see other sources saying prices start at £120, but that excludes VAT and postage).

The real power of the Atom, though, was that it brought Acorn to the attention of the BBC. A British company that could create computers suitable for home users, wrote its own BASIC, and used exactly the type of keyboard the BBC felt was essential? The choice was surely elementary.

Mighty Micro effect?

One popular but incorrect story about the BBC Micro is that it all happened because of an ITV series of documentaries called *The Mighty Micro*. Broadcast in October 1979, and written and presented by Dr Christopher Evans, this took a different approach to the BBC shows of the time.

In particular, it had a philosophical rather than practical tone, as is reflected by episode titles such as 'The coming of the Microprocessor', 'The Intelligent Machine', and 'All our tomorrows'. The thrust of the series was how the coming microcomputer revolution would affect our democracies, the way we communicated, and our working life.

While there's no doubt that *The Mighty Micro* was influential, it post-dates the BBC's Horizon documentary *Now The Chips Are Down* in 1978. It was this provocative programme that prompted a government body to give the BBC £10,000 to research what was happening in other countries.

According to both John Radcliffe and David Allen [9], this led to the idea of a big, multifaceted project that would educate the British population on what computers could do. And the rest is British computing history.

What came next

BBC Model B+64, Model B+128

Release 1985   **Price** £470 for 64kB model, £500 for 128kB model

As the name gives away, the big upgrade here was to memory, with Acorn releasing 64kB and 128kB versions of the Model B+. The processor enjoyed a mild upgrade to the 6512A, but still at 2MHz, while it also gained a floppy disk controller (the plain Model A and Model B had an optional floppy interface).

BBC Master Series

Release 1986   **Price** £499 for Master 128

While it would never be the monster hit of the Model B, the BBC Master was a substantial upgrade on the original series. Above the all-new numerical keyboard, two cartridge slots provided easy ways to add software. It also featured the fourth iteration of BBC BASIC, called BASIC IV.

At launch, Acorn released five variations on the Master. For example, the £624 Turbo added a 4MHz 6502 second processor, while the Master 512 included (obviously) 512kB of RAM, along with an Intel 80186 processor, DR-DOS and GEM software, plus a mouse. This would cost around £1,000.

Acorn would later launch the Master AIV, which included a SCSI interface and a VideoDisc filing system to support the BBC's Domesday Project.

BBC Master Compact

Release 1986   **Price** £440 with no monitor

Acorn's final variant on the BBC Micro was the Master Compact. This offered a separate unit for a 3.5-inch floppy disk drive, and featured a graphical user interface created by Acorn (for the first time). Unlike the main Master series, there was no way to add a coprocessor, but its real problem was its price compared to the Amstrad competition.

What TV programmes came next (and before)

Thanks to a huge archiving effort by the BBC, you can view all the broadcast programmes (and try out the accompanying software) at clp.bbcrewind.co.uk.

The Silicon Factor

3 episodes   March–April 1980

These hour-long episodes were fronted by Bernard Falk and covered how the microprocessor could change the shape of British industry.

Managing the Micro

5 episodes   May–June 1981

Aimed at small businesses, this practical series – consisting of five 25-minute programmes presented by Brian Redhead – focused on how computers were already being used by forward-thinking companies.

The Computer Programme

10 episodes   January–March 1982

The series that gave birth to the BBC Micro, these information-packed (and, on occasion, quite entertaining) episodes all included a practical element of programming along with more general-interest pieces.

Making the Most of the Micro

10 episodes   January-March 1983

Building on the practical elements of *The Computer Programme*, these 24-minute shows examined how people used computers and gave viewers more advanced projects to try.

Making the Most of the Micro Live

2 series   **2 episodes**   October 1983 and June 1984

The first two-hour special included live demonstrations of hardware and software, while Kenneth Baker launched the BBC National Schools Software Competition. In the second broadcast, which lasted an hour, Ian McNaught-Davis introduced the idea of hackers and tackled computer-generated art. He also announced the winners of the previous year's competition.

Micro Live

3 series   **46 episodes**   October 1984 to March 1987

Following the success of the previous live episodes, the BBC embraced the live format with three popular series of 30-minute live broadcasts featuring Lesley Judd, Ian McNaught-Davis, and Fred Harris.

The Trojan Mouse

1 episode   5 April 1992

This one-off programme celebrated a decade of the BBC Computer Literacy Project, tracing the growth of computers from do-it-yourself-kits to home computers used 'for games and by boys'. It also showed how they had been used in schools and beyond.

Sources

Interviews with Christopher Curry, Mike Fischer, Steve Furber, Hermann Hauser, Andy Hopper, Dick Pountain, Chris Turner, and Sophie Wilson.

1. Martin Hayman, Clive Sinclair interview, Practical Computing, July 1982
   worldofspectrum.org/CliveSinclairInterview1982
2. David Allen and Steve Lowry, YouTube, The BBC Micro and Computer Literacy Project
   youtu.be/R7feFEWu25A
3. *Towards Computer Literacy*, The BBC Computer Literary Project 1979-1983
   computer-literacy-project.pilots.bbcconnectedstudio.co.uk/media/BBC-CLP-Towards-Computer-Literacy.pdf
4. John Radcliffe (Executive Producer), The BBC Computer Literacy Project Interviews
   computer-literacy-project.pilots.bbcconnectedstudio.co.uk/aef4287b5161099b147d9c4a75f9c1d3

5. David Allen and Steve Lowry, YouTube, The BBC Micro and Computer Literacy Project
    youtu.be/R7feFEWu25A

6. David Kitson (Head of Transmission Group within BBC Designs Department), The BBC Computer Literacy
    Project Interviews
    computer-literacy-project.pilots.bbcconnectedstudio.co.uk/2059826efbe8f45f25309cc5a6d7528a

7. Kenneth Baker, Reflections with Peter Hennessy, Radio 4, Series 1, Episode 4, first broadcast
    23 August 2016
    computer-literacy-project.pilots.bbcconnectedstudio.co.uk/32f106f530306af5ba52b51c59333fd5

8. Charles Moir, BBC Micro review, Practical Computing, January 1982, page 57

9. John Radcliffe (Executive Producer), The BBC Computer Literacy Project Interviews
    computer-literacy-project.pilots.bbcconnectedstudio.co.uk/aef4287b5161099b147d9c4a75f9c1d3

ZX Spectrum

Not the BBC Micro

When Clive Sinclair stepped onto the stage at London's Churchill Hotel to announce the ZX Spectrum, it had been over a year since the BBC had awarded Acorn the contract to build the Micro. The wounds were still fresh. 'At one point we thought of calling it "Not the BBC Micro",' he declared [1], a verbal flourish of the dagger before the attack: 'It's obvious at a glance that the design of the Spectrum is more elegant. What may not be so obvious is that it also provides more power.' [2]

With the BBC Micro metaphorically reeling, he went in for the kill with detail after detail. 'The ZX Spectrum has more usable RAM and higher maximum RAM. It offers twice as many colours on the screen at any one time, plus a colour brightness control. It also offers definable graphics. It has a data transfer rate 25 percent faster, supported by a VERIFY facility. And it employs a dialect of BASIC already in use in over 400,000 computers worldwide.'

He would end on a stinger. 'We believe the BBC makes the best TV programmes – and that Sinclair makes the world's best computers!' Take that, BBC Micro.

The watching journalists, always keen for a show, lapped it up. And they liked the Spectrum even more once they got to try it for themselves: this was a computer you could buy for £125 and that did, in all the ways that Sinclair outlined, offer much more than the £299 BBC Micro Model A.

While Clive soaked up the glory and the headlines, the truth was that he had little involvement with the development of the Spectrum. Just as with the ZX80 and ZX81, computers didn't grab his imagination in the same way as the Sinclair C5 electric car and the miniature, portable TV. Microcomputers were the money-makers, but he had previously handed the computer-making reins to his most trusted and loyal engineer, Jim Westwood.

This time, there was a changing of the guard: he wanted Westwood to concentrate on the problems besetting the miniature TV project, so the responsibility of creating the 'ZX82', as the project was codenamed, fell to Richard Altwasser. With no clear direction from Clive Sinclair – other than to keep it cheap and deliver the project by spring 1982 – Altwasser and other engineers at Sinclair brainstormed their way to a working specification: 'We decided it must have high-resolution graphics, probably 16K of memory, an improved cassette interface, sound, and of course most importantly colour.' [3]

They quickly realised that the ZX81's approach to display output – which dedicated three-quarters of the Z80's processor time to running the display – was not going to satisfy people buying a high-resolution colour computer. They would 'want to have fast-moving animated displays,' says Altwasser. 'Consequently, we decided to design a computer architecture that divorced the CPU from the display.'

Nothing too revolutionary there. What was revolutionary, and soon became a trademark of graphics on the Speccy, was the engineers' approach to colour. Most of the ZX Spectrum's colour rivals included a dedicated video chip, but with no budget for this they came up with an ingenious solution. Rather than imbue each on-screen dot with its own colour, they split the screen up into 8×8 grids called 'attribute blocks'. The Spectrum had a resolution of 256×192, which equates to 32×24 attribute blocks.

Each block could only support two colours at a time, with a total of eight to choose from: black, blue, red, magenta, cyan, green, yellow, and white. Adding a 'bright' attribute to a block meant you could cheat your way to 15 colours (in case you're wondering what happened to unlucky number 16, black looked the same either way).

So now you have the question, in an 8×8 grid, how does the display know which is, say, red and which is blue? The answer comes in the Speccy's INK and PAPER commands. Take the 8×8 grid below, spelling out the lower-case letter 'a'. The first column would all be zeroes to indicate paper, while the second column has one square that is a one to show that it's ink.

This skinflint approach to colouring in the dots meant the Spectrum only used 7kB of its memory for display output, giving programmers a minimum of 9kB to play with (assuming people were using the Spectrum 16K). And if you're wondering how the Speccy supported flashing, that could also be attributed at the block level. 'The whole 8×8 grid flashed by swapping ink and paper colours for 0.64 seconds, but one dot couldn't flash on its own,' says Rupert Goodwins, who joined Sinclair Research as a software engineer in January 1985 – the same week that Sir Clive launched the C5.

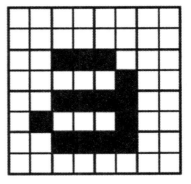

Image taken from the ZX BASIC manual [4]

On to the next challenge: improving the tape interface. This was necessary because programs could go up to 41kB in size, and the current tape technology operated at around 250 baud – that is, 250 bits per second, or just over 31 bytes per second. That means a 1kB program would take 33 seconds to load, while a 41kB program would keep you waiting 20 minutes. Even in the 1980s, when they had just four TV channels to choose from, British kids would have found something else to entertain them by then.

'We originally aimed at getting [the tape mechanism] to work at about 1,000 baud, but we succeeded in making it work at 1,500 baud, which is significantly faster,' says Altwasser.

Many histories of the Spectrum suggest that it included a piezo-electric beeper rather than a speaker, but don't tell that to Goodwins. 'No! This myth has been repeated so often I just had to go and check one of my own stock of Spectrum PCBs,' he says. 'The Spectrum has a proper moving-coil dynamic speaker. A very small one, but absolutely a real speaker. It can only be turned on and off, with the BEEP command just sending a string of pulses at a certain frequency for a certain time, but with clever programming it can produce multi-voice music, sound effects, and even barely recognisable speech.'

Few programmers were clever enough to take advantage, however, with most games defaulting to a simple 'bleep'. This seeming inadequacy was to be a frequent jibe aimed at the Spectrum, especially when advanced audio machines such as the Commodore Amiga emerged, but at the time Sinclair fans were simply grateful to hear anything at all.

If Sinclair had left the hardware upgrades at that, the Spectrum would probably have been a success but certainly not the gargantuan hit it was. But it also had style. A significant amount of praise for this should go to the late Rick Dickinson, whose pride in his work shone through in an interview with Sinclair User [5]. 'I like the Spectrum much more than the ZX81… It is a step upmarket and I was really trying hard for a super-smart machine. It is not for quite the same amateur market.'

While Dickinson didn't add a second Design Council award to his mantelpiece, having earned one for the ZX81 the previous year, the new computer's design was widely praised by reviewers of the time. 'The Spectrum is a smart, slimline machine,' declared Popular Computing Weekly [6]; 'It looks extremely elegant,'

Personal Computer World [7]; 'the Spectrum comes in a smart little black box', Practical Computing [8].

And it's fair to say that the Spectrum's design has stood the test of time. Eurogamer's Ian Higton put it well in the site's obituary to Rick Dickinson, who died in 2018, when he described the Spectrum as uniquely 'loveable' [9]. Who could fall in love with the statuesque BBC Micro or simply dull Commodore 64? The rainbow-splashed Spectrum, so easy to throw into a rucksack, was the opposite: quirky, friendly, fun.

After repeated criticism of the membrane keyboard on the ZX80 and ZX81, however, Dickinson knew that he had to come up with something that actually moved this time. 'We spent a great deal of time on that,' he said. 'It is the only interface between the user and the product and it has to be right.'

As with the Speccy's sound output, not everyone was a fan of the keyboard's 'dead flesh' feel, but John Grant of Nine Tiles, the company that (as with the ZX80 and ZX81) created the operating software and BASIC for the Spectrum, still defended this decision in a 2018 interview. 'I think it was the best that one could do with the technology that was available at the time,' he said [10]. 'I think it worked pretty well, and anything that worked better would have been a lot more expensive.'

Sinclair also decided to stick with its approach of printing BASIC commands onto the keyboard. So, where you would type out the word 'INPUT' on a BBC Micro or Commodore 64, with the Spectrum you needed to locate it on the keyboard – in this example, it's on the 'I' key. For the command 'IN', you pressed the SYMBOL SHIFT button and then the 'I' key – 'IN' was printed below the key in red letters.

To an extent, this was useful. Not everyone is a touch typist, and when you're copying a program from magazine listings – as any geek who grew up in the 1980s will know, this was a common activity – there were plenty of occasions when it saved time. Those budding programmers didn't know it, but they should also have been saying silent thanks to Nine Tiles' John Grant. 'I put quite a lot of effort into the decoding of it so that you wouldn't get false key presses, double key presses, or seeing two key presses as one,' he said.

The main architect of the Spectrum's hardware is adamant that most of the credit for the Spectrum's BASIC, however, should go to Steve Vickers. 'He pretty much single-handedly wrote, tested, and debugged the Spectrum BASIC, and wrote the instruction manual,' Altwasser says.

Spectrum BASIC added more functions to its predecessors, which meant yet more keywords to go with each key. As a rule, these had five distinct meanings depending on the mode you were in, so you needed to locate the function, switch to the mode associated with it, and then press the key. For some three-letter keywords, such as INK, this meant just as many button presses as typing in the word itself.

While this was confusing to many Spectrum users, compensation came in the form of a much-improved approach to saving your work. Whereas on the ZX80 and ZX81 this was a case of cross your fingers and hope, especially if you were relying on the 16kB memory expansion pack not falling out, a VERIFY command in the Spectrum meant you could save your program to tape and then check it matched what was in the computer's memory. For anyone who had spent hours typing in a long program only to discover it hadn't saved to tape properly, this was worth the £125 asking price all on its own.

Grant cleverly included a video show for children to watch while waiting for programs to load. While the Spectrum's processor was 100% dedicated to decoding the data streaming in from the tape, he worked out a way to put 'coloured stripes on the screen that showed you how it was decoding... So you could actually see on the screen, and you could hear the warbling, how the process was going. Which I think was helpful.'

Even today, Rupert Goodwins remains impressed by Grant's trick. 'It actually flashes the colour of the border of the screen fast enough to make it stripy (the border is normally set to its own colour from BASIC with the BORDER command),' he says. 'The main screen is untouched, although graphics could also appear on the screen proper during loading, building up in monochrome line by interlaced line, with the attributes coming in last and colouring the screen in one quick wipe. It's quite charming.'

Ultimately, though, it's software rather than charm that sells computers, and while a tape designed for the ZX80 or ZX81 wouldn't work with the Spectrum – the different baud speeds meant they were incompatible – the popularity of the Sinclair brand, the similarity between the ZX81's internals and the Spectrum's, and a booming number of British games developers meant the Spectrum rapidly developed a games catalogue the envy of the computing world.

Highlights? Even while Sinclair was struggling to get enough Spectrums out of its factory doors, Melbourne House had produced a version of *The Hobbit* for the Spectrum 48K. *Manic Miner* was another early and long-lasting hit from Matthew Smith, who cemented his position as a gaming legend by following it up with *Jet Set Willy*. Then there was *Chuckie Egg*, created by British teenager Nigel Alderton. The list could go on and on.

As with the ZX81 before it, Sinclair's biggest problem was meeting demand. Ignoring the usual issues of squeezing a reliable ULA (Uncommitted Logic Array) out of Ferranti – a common theme for the British-made computers in this book – Sinclair Research suffered a number of production-related blows during 1982.

Perhaps the biggest was the departure of Altwasser, who left – taking Steve Vickers with him – in early 1982, soon after production of the Spectrum began. They would go on to produce the ambitious but ill-fated Jupiter Ace, which shared many traits with the ZX Spectrum but used Forth rather than BASIC. On a minor scale, this replays the familiar 'superior doesn't mean successful' pattern: while Forth is widely recognised as being a far superior programming language to BASIC, the British public wasn't ready to switch. And without numbers, developers won't produce games. And without games, computers didn't sell.

But back to the Spectrum. Nine Tiles' John Grant would later complain to Sinclair about a lack of structure in the project, with the software completed in April 1982 but no hardware to test it on. In particular, they had left 1,300 bytes free to include the software to control peripherals as they had no hardware to control.

Despite this, with an expectant public to feed, Clive Sinclair took the decision to start production of the Spectrum. The idea, according to an interview with John Grant in 1985 [11], was to produce a few Spectrums and then replace their ROM chips – read-only-memory chips, which stored the operating system – when they were finished. 'Then it got to somewhere around May or June and they'd sold 75,000 machines, all with the old ROM. They came to the conclusion that the original idea just wasn't viable.'

The answer, such that it was, came in the form of an add-in card that would act as a 'shadow ROM'. This kicked into action whenever a peripheral needed to be used, and while it worked it gained zero points for elegance.

Meanwhile, Sinclair and Nine Tiles' relationship – which had started with the ZX80 in 1980 – was coming to a sudden end. 'I went to see Clive in Portugal Place and Clive said our hardware suppliers are giving us lower prices now, so we want lower prices for the software,' Grant recalled in the *Floppy Days* podcast. 'And I said, but your hardware suppliers are giving you low prices because you're buying more hardware. Whereas you're not going to be buying more of our time. So no, I'm not going to reduce the price per man hour. However, if you want us to be collecting a royalty per chip, then, yeah, we can talk about that. And Clive said, I'm not telling you how many I sell. So, that's kind of where it all stopped.'

The third act in the drama-farce that was the Spectrum's early production came through no fault of Sinclair's but due to problems at the Timex plant in Dundee. This huge facility employed over 4,000 people, but when it lost a huge contract the factory made many workers redundant. A factory-wide strike was followed by the plant's scheduled three-week summer holiday, and all the while the orders were stacking up.

With Altwasser gone, the responsibility for solving production problems fell on the young shoulders of John Mathieson (although Jim Westwood was also called in to help). Mathieson would go on to co-found his own company before joining Atari, VM Labs, and Nvidia in a variety of senior positions, but in 1982 he was a fresh-faced Cambridge University graduate. 'I did a lot of work to redesign the mainboard to make it more manufacturable and more reliable,' he says. 'The Spectrum was very hard to manufacture so it was difficult to get volume going.'

Take the vital aspect of a colour computer: outputting colour to the TV. 'When you drive a colour television, the colour information is carried separately from the black and white data,' says Mathieson. 'It's how they added colour to black and white TVs. And there's a very specific carrier frequency for it. So we had a crystal in the Spectrum that generated that carrier frequency: unless it's set very accurately, you don't get colour. That was one of the calibrations they were doing on the production line, and it was a little variable capacitor. These things would get vibrated on the production line and during shipping and drift out of their original setting. And then the colour no longer worked.'

With low yields, the factory had no choice but to shut down the production line and call down to Cambridge. 'So we would jump on the Sinclair company plane and fly to Dundee, debug a lot of broken computers, and then get the production line going again.' Mathieson estimates that it took a year to get rid of all the questionable design decisions, whether that was redesigning the circuits to generate colour or replacing a power supply transistor that was underspecified and had the undesirable habit of blowing up.

At one point, those who placed orders for the Spectrum – which, initially, were through mail order only – faced a waiting time of twelve weeks. Sinclair published an apology and the offer of £10 towards the ZX Printer, while slashing prices of the ZX81 in the hope of attracting some customers to last year's machine. Still, though, it continued to advertise the Spectrum in the computer press.

With a backlog of 40,000 machines by September 1982, orders continuing to flood in, and a production run of about 5,000 Spectrums per week, Sinclair was facing the usual problems of a successful British computer manufacturer in the early 1980s: demand far outstripping supply. No wonder that many British kids who wanted a Spectrum under their 1982 Christmas tree ended up with a Dragon 32 or cut-price Commodore VIC-20 instead.

It didn't help that Sinclair kept on advertising the Spectrum when it knew that it couldn't keep up with demand: 'Production delays with the Spectrum resulted in the Advertising Standards Authority receiving a record number of complaints,' read the January 1983 edition of Sinclair User [12]. 'The authority noted that Sinclair Research failed to withdraw the advertisements when it became clear there were supply problems.'

Once Mathieson and his colleagues had ironed out the manufacturing problems, though, there was no stopping the Spectrum in 1983. With production running smoothly, Sinclair struck with a dramatic price cut: the Spectrum 16K now cost £99.95, the 48K £129.95. This immediately made both the Dragon 32 (£199.50) and the VIC-20 (£139.50) look overpriced, while it was even more difficult to justify £345 for the Commodore 64. Meanwhile there was no sign of the promised Acorn Electron, still stuck in development hell.

The Spectrum had another big factor in its favour: software. By spring of 1983, WHSmith was taking out double-page adverts in Sinclair User that offered 65 programs, from Psion's *Flight Simulator* to the legendary *Horace Goes Skiing*, from Address Manager to Vu-Calc.

Clive Sinclair was also starting to harvest the fruit of his long-held ambition to position the Spectrum as a schools computer (and stick it to the BBC in the process). The first seed was planted in July 1982, when prime minister Margaret Thatcher announced that primary schools would now benefit from the government-sponsored computer buying programme and that the Spectrum was on the approved list: this meant the government would subsidise 50% of the cost of the computer. Clive Sinclair quickly used his usual marketing nous to take full advantage. He promised all primary schools – then numbering 27,000 – a free ZX Printer worth £59.95 if they bought a Spectrum rather than the other machines on the list, namely the BBC Micro Model B and Research Machines' 480Z.

A year later, Sinclair took out full-page adverts in the daily newspapers with the headline, 'Popular in schools. And the only computer that runs primary schools software at home.' [13] The message was a spin on a now familiar formula: if you want the best for your children, you need to buy them a computer. 'The trouble is, that even though the [school] computers are subsidised, there are likely to be more children than computers… The solution, of course, is to buy one of the approved computers and carry on the good work at home. By far the cheapest of these computers is the ZX Spectrum.'

In truth, and in stark contrast to the opening line of the above advert, the Spectrum never made much of an impact in schools. At its highest point, it represented 2% of computer sales to schools, with the vast majority still choosing the BBC Micro Model B. Unless you were called Clive Sinclair, it was obvious why: the BBC machine was built to take a beating from school kids while the Spectrum struggled to last a school term.

The problem, explains John Mathieson, who would go on to manage Sinclair's support team for the Spectrum, wasn't the keyboard but the frame. 'There was a metal cover over it and it was just held down with double-sided sticky tape. And that sticky tape would fail after a while, especially with the heat coming up from the computer.' Little wonder, then, that the Spectrum was almost as renowned for its failure rate as it was for its gaming prowess.

Mathieson was also fighting a continual battle to improve the reliability of the Spectrum, bringing out several revisions of the main board throughout its life. Not all those improvements went as planned, however: if you bought an 'issue three'

Spectrum in the middle of 1983, then there was a good chance some commercial tapes wouldn't load. Bob Denton, managing director of Prism Technology Holdings, distributor of the computer to many high-street retailers, said 'The number of returns is horrific.' [14] He estimated that two in five 'issue three' Spectrums were being returned.

This inevitably caused a second hiccup in supplies, with Sinclair User publishing this comment from a Sinclair spokesperson in the July 1984 issue: 'The shortage is due to an incredible sustained demand for Spectrums... There are Spectrums around, it is unfortunate if they are not available in some areas.' [15]

The saga of the Microdrive did little to enhance Sinclair Research's reputation for reliability – or sticking to its delivery time frames. With the heavy application of what Mathieson calls 'smoke and mirrors', this had been announced with great fanfare at the launch of the Spectrum. 'Clive wanted to pretend it was pretty much done, so there were these little things on top of a bench, but there was an awful lot of electronics hidden underneath.'

The watching press was both hoodwinked and impressed. 'Perhaps the biggest rabbit that Clive pulled out of his magician's hat was the ZX Microdrive,' wrote Nick Hampshire in Popular Computing Weekly, as part of his review of the ZX Spectrum [16]. 'This is a very tiny disk drive, using [2.25-inch] diskettes, with each diskette capable of holding up to 100K bytes, and a transfer rate of 16K bytes per second.'

The Observer newspaper was similarly upbeat, describing the Microdrive as 'astonishingly cheap' [17].

Other disk drives stored more data, but the computer-buying public was conditioned into paying hundreds of pounds in return. Sinclair, on the other hand, was promising the Microdrive would only cost £50. And it would be available later that year, too.

At that launch event, he was deliberately coy about the technology inside. This left the computer press to clutch at straws, with the key one being the phrase 'a single interchangeable microfloppy' in adverts that appeared in the summer of 1982 [18]; many believed the Microdrive would be a diskette similar to the Sony 3.5-inch floppy drive that was due to appear that year for around £250. By October, though, all mention of microfloppy had been removed from Sinclair's ads.

In reality, the technology inside the ZX Microdrive was more akin to the continual loop tape drive found in that era's answering machines. The key concept was to store data on a five-metre loop of magnetic tape that could swoop round in eight seconds: the Microdrive would spot the relevant section of data during one of these loops and then transmit it to the Spectrum.

Despite some heavily gilded claims from Clive Sinclair that this was a Sinclair Research invention, the concept wasn't new. 'I have heard Clive claim that it was a totally original idea,' says former Sinclair MD Nigel Searle, 'but I know that when I was in the States, he asked me if I could order and ship over to him one of these [Exatron] drives.' Indeed, to anyone who understands how the ZX Microdrive works, the Exatron Stringy Floppy will sound familiar: 'to load a specific file the drive searches the entire tape, briefly stopping to read the header of each found file' [19].

As ever, though, the Sinclair difference was in size and price. The Exatron drive cost $249.50 and was roughly the size of a cassette recorder, while you could hold a Microdrive in the palm of your hand. The promise of affordable, fast storage – fast compared to cassette recorders, at least – seemed like a dream.

Behind the scenes, those involved with the Microdrive project were having something of a nightmare. Clive Sinclair's prediction that the product would launch by the end of 1982 proved to be a year wide of the mark, despite the valiant efforts of Ben Cheese, John Gilbert, and Martin Brennan. While Cheese never spoke publicly about the development of the Microdrive, and sadly died from cancer at the age of 41, we can only imagine the difficulties in creating the ULA in partnership with Ferranti to make it work.

In typical form, Sinclair Research decided to launch the ZX Microdrive in late 1983, before its production wobbles had been ironed out. A month after it reviewed the Microdrive – where it was praised for its simplicity, low cost, and reliability – Sinclair User featured an interview with the creator of the Microdrive ROM, Dr Ian Logan, who explained that the software wasn't finished and still contained known bugs. 'When you can get a Microdrive with a nicely-covered ULA and tidily-set-out board you will know that Sinclair is finally convinced that everything is right,' he said [20].

Sinclair's marketing also glossed over the fact that you couldn't simply plug the Microdrive into your Spectrum: you had to buy the £29.95 Interface 1 module and then attach the Microdrive to it. Consolation came in the form of local networking ability and a serial RS-232 port, which rapidly led to a flourishing market of Spectrum-compatible printers. (Sinclair's own ZX Printer was quietly discontinued in the middle of 1984.)

Two other problems affected the Microdrive's popularity at the start of its life. First was its extremely limited supply, with fewer than a thousand shipped after three months. Indeed, buying one was on an invite-only basis at first. Then there was the price of the cartridges themselves, which cost £5 compared to around £2.50 for a floppy disk at the time. While the price of cartridges soon dropped to £2, the Microdrive's lack of sales and high cost of production (compared to tape) meant that few software developers ever made the switch to the new technology.

Despite this, Charles Cotton – who spent three eventful years as sales and marketing director at Sinclair between 1983 and 1986 – looks back on the Microdrive with fondness. 'Initially with the Spectrum it was a very good solution for expanding the memory as a kind of hard drive,' he says. 'It was not quite as successful when it got to the QL because it was at 85K and we tried to expand the memory to 100K, and that was probably a step too far.' As we explore in the story of the Sinclair QL (page 212), it's hard to argue with that last statement.

The ZX Interface 2 followed hot on the heels of the Interface 1, but this was an out-and-out flop. When viewed from four decades' distance, you wonder why Sinclair Research even produced it. No doubt the biggest reason was to add joystick ports to the Spectrum, an omission that had long been criticised, but the third-party Kempston interface had filled this void. And the ZX Interface 2 wasn't compatible with Kempston, meaning games that used joysticks made before the Interface 2 didn't work.

The other inclusion in the Interface 2, and one that would be easier to defend aside from one vital flaw, was a ROM cartridge slot. If third-party manufacturers had embraced this, it would have meant games loading instantly. But each cartridge could only hold 16kB of data – the vital flaw we speak of – when most games were targeted at people who owned the more popular Spectrum 48K. According to Wikipedia, only ten games were ever released on Spectrum ROM cartridges [21].

As is the case with so many of the British computers of the 1980s, the problem stems from timing. All those peripherals emerged over a year after the Spectrum itself, by which time third-party companies had stepped in to fill the breach and make money.

And, while Sinclair did have success by publishing third-party software under its own brand, this was never a core part of its strategy. 'Clive was very uninterested in software,' says Sinclair managing director of the time, Nigel Searle. 'But really it was what drove the sales of the Spectrum.'

What ultimately caused the problems for Sinclair, however, was reliability. This was always a blind spot for Sir Clive – he was knighted in the Queen's 1983 Birthday Honours – who consistently railed against any accusation that Sinclair products were unreliable. This thorny subject was the cause of his much-publicised fisticuffs with Christopher Curry in December 1984, after Acorn placed a full-page advert that listed ZX Spectrum failure rates at 24% compared to 5% for the BBC Micro.

To quote a barman at the scene, a pub in Cambridge called the Baron of Beef, where both Acorn and Sinclair employees were regular drinkers, 'Sir Clive was obviously furious about the content of an Acorn advertisement, a copy of which he had in his hand. There was some very strong language indeed.' [22] By all accounts, though, the two quickly restored normal relations, and remain friends to this day.

What may have hurt, though, was the truth behind the claims. 'Ultimately what brought Sinclair down was the return rate on [Spectrums] was somewhere around 20%,' says Mathieson. 'If your sales keep growing rapidly, it's a burden you can manage because by the time they come back, your sales will be growing so you remain profitable. But as soon as it plateaued, it kind of caught up with everything.'

Another problem was Sinclair's generous returns policy, where its users could return computers within a year. That was workable when your users are hobbyists who will look after their ZX81, but it's a different story when those users are children. 'We had a rather silly policy that if you took your computer back to the place where you bought it from – many of them were sold through Smiths for example – you could get a new one,' says Cotton. 'And so little boys were taking their computers back and getting a brand new one when there was nothing wrong with them at all. They maybe were getting dirty or whatever else, but we ended up with a very large number of so-called non-working or broken Spectrums. I mean, 50,000 is a huge number.' Multiply 50,000 by the £50 Cotton estimates it cost Sinclair per replaced Spectrum, and you rapidly come to £2.5 million.

Clive Sinclair echoed this view at the time. 'Many of our customers, in fact the majority of them, are in the 14- to 15-year-old age bracket,' he said [23]. 'These are characters who can destroy granite with one blow of their fist, so a computer gets a pretty tough pounding over the course of a year – Coca-Cola spilt on it, this sort of thing… So yes, we get a lot back, but they aren't faulty computers, they're ones that have got damaged and so on.'

Slow sales during Christmas 1984 – an industry-wide problem that almost wiped out Acorn due to poor sales of the Electron – only made Sinclair's financial woes worse. The world, slowly, was also moving away from 8-bit computers and towards the bright new future heralded by the Apple Macintosh: a world of powerful 16-bit processors, graphic user interfaces (GUIs), and more versatile software.

Despite these challenges, Spectrum sales spluttered on through 1985. Its most valiant attempt to breathe new life into the platform was the Spectrum+, with a proper keyboard borrowed from the Sinclair QL. 'We eventually got one more year of very good sales out of the Spectrum by putting essentially the QL keyboard technology onto the Spectrum,' says Searle. 'We sold a lot of those.' Unfortunately it also suffered from reliability problems, with one claim that Boots (a key retailer for the Spectrum) saw return rates of 30% [24].

Then there was the problem of manufacturing rejects. In 1985, having seen the Sinclair QL through its difficult birth, David Karlin took on the head of manufacturing role at the company. 'The main thing to be done was the serious process of refurbishing the world's largest pile of reject Spectrum boards,' he says. 'There were piles of tens of thousands of these things sitting at Timex and Thorn-EMI Data Tech, and so during my last days at Sinclair I was spending my life basically turning those into shippable product.'

Speaking to Nigel Searle, it seems almost inevitable that Sinclair's computer adventure would end like this. 'Part of the problem with Clive was that he was never happy unless he was placing very, very big bets. It wasn't exciting to him. It was only exciting if he might lose his shirt. A safe investment? Oh, how very boring. He put everything on a roll of the dice, every time.'

In June 1985, news came through that a company owned by Robert Maxwell would be paying £12 million for a 75% stake in Sinclair Research, leaving Sinclair with 20% and other shareholders the remaining 5%. 'I am not the sort of person to run an established business,' said Sir Clive, quoted in The Times [25]. 'I am good during the early up-rush, then it needs other hands... I am an inventor.'

Within weeks, though, this deal had collapsed, only for Sinclair to announce Dixons would be its new saviour. 'Within the last few days we have concluded an immensely exciting contract with the Dixons group which is worth in excess of £10 million to Sinclair Research over the next three months,' he said [26]. 'Our problems were always of a short-term nature and whilst we were grateful to Bob Maxwell for his support, we are happy to be continuing as an independent company.'

This wasn't the truth, the whole truth, and nothing but the truth. It soon emerged that Maxwell expected Sir Clive to stay out of the computer market for five years as part of the agreement, while Sinclair thought he had the freedom to start afresh. Maxwell's financial advisors, on going through the final numbers, appear to have poured cold water on the deal too.

By Christmas 1985, all seemed rosy once more. Sales had been better than expected, thanks in part to the launch of the Spectrum 128 in Spain; with this product set for sale in the UK early in 1986, there was reason for optimism.

It was with some surprise, then, that people read this headline on the front page of The Guardian on 8 April 1986: 'Sinclair £20m debts force sale to rival' [27]. The opening paragraph made grim reading for his employees: 'Sir Clive Sinclair is selling his company's home computer assets, including rights to use the world-famous Sinclair brand name, to Amstrad Electronics, his more successful, larger, and highly profitable rival.' Ouch.

Despite Sinclair's optimistic statements to the contrary, this had been a long time coming. His company owed money to Timex and Thorn-EMI, while a huge payment was due to Barclays Bank on 31 March 1986. Price Waterhouse had been commissioned to find a buyer and approached the Dixons Group; they weren't interested, but knew someone who might be. Cue Alan Sugar.

'On my second call with the chap [from Price Waterhouse], it became clear to me that there was a deal to be done,' wrote Sugar in his autobiography, *Alan Sugar: What You See Is What You Get* [28]. 'Now, here is where I defied all business logic. With no deal done, I decided there and then – before meeting Clive Sinclair or discussing numbers with banks – that I was going to buy the Sinclair business one way or another.'

No matter what you think of Alan Sugar, it's impossible not to admire what he did next.

First, and still not having met Sinclair, he sketched out a design with his right-hand man Bob Watkins for what would become the ZX Spectrum +2: essentially the Spectrum 128, so with a proper keyboard, and an integrated cassette recorder. (The Microdrive was history.) He commissioned a prototype and got his Hong Kong office to draught a detailed model. Knowing that he needed to act quickly if he was to bring out a new product in time for Christmas 1986, he then instructed his preferred manufacturer in Taiwan to buy the tooling for it, at a cost of around $100,000.

Still not having met Sinclair, he showed Dixons the prototype and convinced them to put in an order for 100,000 units with an agreed retail price of £139.99.

Sugar finally met Sinclair the following week and explained what he was willing to do – and, just as importantly, what he wasn't. At this point, according to Sugar, Sir Clive still hoped to run the business; he basically wanted Sugar to bankroll him. By contrast, Alan Sugar was after the rights to manufacture Sinclair-branded computer products. A reluctant Clive Sinclair eventually agreed, but by this time he had no real choice: at the end of March, Barclays Bank would otherwise foreclose his business.

Price Waterhouse arranged a meeting on 24 March for all the affected parties. These included Sinclair's manufacturers Timex and AB Electronics, both of which held stock and were busy making new computers. They wanted payment or the deal wouldn't happen. Sugar's pragmatic solution was to agree a price per unit, and then offer that to Dixons at cost price. Done. Next?

This left Barclays Bank as the only unhappy party, and it wouldn't sign the deal. Sugar had a plane to catch so left it with an ultimatum: this was his deal, take it or leave it. You have seven days to decide. On 30 March, after a marathon all-day meeting in London where details of proposed deals were faxed to Sugar for his approval, they finally reached an agreement.

Except for one thing. Sugar wanted all the rights, and that included Nine Tiles' work on the operating system. Due to the impasse described earlier about fees, Nine Tiles owner John Grant had never signed any documentation to say that Sinclair owned the operating system. We will let Grant take up the story.

'We'd gone off for Easter on our canal boat with family and of course this was before the days of mobile phones so there was no actual way to contact us to say, please we want you to sign this across.' Cue a frantic search for him by Jim Westwood. 'We had told some people that we were going on the canal and we were going somewhere with a flight of so many locks or whatever. And he worked out from that roughly where we ought to be and actually went over there and looked for us but didn't find us. We worked out later that at the point where he gave up if he'd gone on and gone around the next corner, he would have found us.'

Fortunately, Grant returned home on the evening of Easter Monday and noticed his answerphone flashing. The messages were rather desperate: if he didn't sign across the rights, the Amstrad deal wouldn't go through and Sinclair would go bust. Cue another frantic round of calls, this time from Grant to his lawyers to see how much he could sell the rights for. The answer: not much. In double-quick time, Grant and Amstrad agreed a price of £20,000 for the Spectrum rights and the deal squeezed through.

Sugar claims that they sold 300,000 units of the ZX Spectrum+ 2 that Christmas, and while it lacked the elegance of the original Spectrum it was far more practical. With 128kB of RAM, a spring-loaded keyboard, dual joystick ports, and that built-in cassette recorder, it was a tempting upgrade choice for existing Spectrum owners too.

Thanks to Amstrad's intervention, Spectrum computers kept on selling into the early 1990s, bringing the estimated total number of sales to 5.5 million. While many of those were sold abroad, the Spectrum was undoubtedly the most popular home computer in Britain during the 1980s: if you didn't own one yourself, chances are that a friend of yours did.

As such, its impact on the UK is unmeasurable, but certainly on a par with the BBC Micro. 'I think there was a generation that grew up with [the Spectrum] and really learned something that no generation before or since did,' says Mathieson. 'The programming was right there: everybody played a little bit with the BASIC, even if they were mostly playing games on it. And so I think there's a generation of game writers in the UK that grew up with the Spectrum.'

For Charles Cotton, it wasn't just the Spectrum that made a difference but Clive Sinclair himself. 'Clive was inspirational for people and I think the Spectrum was inspirational for people as well,' he says. 'I think [they] encouraged many young people to see an entrepreneurial future and it has been a huge boost for the entrepreneurial behaviour in Britain and particularly in Cambridge.'

What's notable in the feedback we received from the British public is the affection they still feel for the Spectrum, and towards Clive Sinclair himself. Terence Thompson described the Spectrum as a major influence, prompting him to learn Sinclair BASIC which 'led to MS-DOS and writing batch files to automate computer-controlled machinery… I eventually changed my job with the same company and worked in the IT department as the development engineer. So thank you very much Clive Sinclair for the ZX Spectrum.'

Mark Walsham found it 'ignited my passion for programming that sustained me, and continues to sustain me, in my profession as an IT consultant 30 years later!' Paul Russell still has his Spectrum safely 'tucked away' and credits it for getting him 'through computer studies at school. I went on to work in computing since 1987… I mostly had a great time and was largely self-taught, which started with my dear old Spectrum.'

The Spectrum 128 Easter egg

The 128 was the only Spectrum to be released abroad before it hit the UK. That's because it was developed with the help of Sinclair's Spanish distributor, Investrónica, which was facing its own version of the BBC Micro vs Sinclair Spectrum battle. Its opposition, though, was a computer based on the Dragon 64, which had supposedly won a broadcasting/computer agreement that mirrored the BBC's with Acorn.

Launched in Spain in September 1985, the Spectrum 128 was based on the Spectrum+ but included 128kB of RAM, an improved audio chip, support for MIDI, an RS-232 serial port, and an RGB output for monitors. Less obvious was that it included an overhauled ROM.

'The original Spectrum source code was in a terrible state,' says Rupert Goodwins, at that point a new hire who had been shunted up from Sinclair's London office to Cambridge to sort it out. 'There weren't any comments. It was just this stream of Z80 mnemonics.' It took Goodwins around two months to recompile it to a state where the Sinclair team, headed up by Steve Berry, could start work adding the expanded support for BASIC, music, and the RS-232 port.

While Berry worked on the much-improved BASIC editor, which would now support typed-in commands rather than just single-key entry, it was Goodwins' idea to add the rainbow stripes on top of the menu bar. 'Marketing got very excited by that.'

You can even find an Easter egg tucked away. 'I wrote a sound-to-light program in my bedroom when I was 17,' says Goodwins. 'In a very few bytes of machine code, you played music into the tape port and it put all these pulsing patterns on the screen in real-time. And I managed to sneak that into the ROM as an Easter egg.' While the Spanish children could enjoy such delights when they opened their gleaming new Spectrum 128 in Christmas 1985, downtrodden British kids would have to wait until it went on sale for £179.95 in January 1986. Annoyingly, this reason was purely down to business: Sinclair didn't want to hurt Christmas sales of the Spectrum+.

What came next

The ZX Spectrum was undoubtedly Sinclair's biggest hit, with two successful follow-ups, and Amstrad took full advantage of the brand with a number of subsequent models.

ZX Spectrum+

Release 1984   **Price** £179.95

In response to market research that showed some people were put off by the original Spectrum's rubber keyboard, and buying the Commodore 64 instead, Sinclair produced the Spectrum+. This was identical to the Spectrum 48K other than its new case and injection-moulded keyboard, both borrowed from the Sinclair QL.

ZX Spectrum 128

Release 1986   **Price** £179.99

With the same case and keyboard as the Spectrum+, the big improvements were inside: 128kB of RAM, a radically improved 32kB ROM (complete with a more advanced BASIC that abandoned the keyword approach of its predecessors), and a sound chip that could output on your TV. There were three notable new outputs too. One was an RGB/composite video out, one for connecting numeric keypads, and one combo proprietary serial/MIDI port (although the MIDI was out only).

ZX Spectrum +2

Release 1986   **Price** £149

The first Amstrad-created Spectrum showed all the hallmarks of Alan Sugar's pile 'em high philosophy, with an integrated tape drive plus the promise of 'bundles' – such as a joystick and six-pack of software for £159. Internally, it was almost identical to the Spectrum 128 aside from improved cooling. This was also the first Spectrum to include dedicated joystick ports.

ZX Spectrum +3

Release 1987   **Price** £199

This replaced the cassette recorder of the Spectrum +2 with Amstrad's favoured 3-inch floppy drive and came with Locomotive Software's +3DOS operating system, but anyone hoping for an upgrade to the Z80 processor (now almost ten years old) would be disappointed. Still, it was compatible with games, and that's arguably what mattered.

ZX Spectrum Next

Release 2020   **Price** £175

Not an official product, but a Kickstarter campaign started by Spectrum enthusiasts in 2017. Safe to say they had production problems, with deliveries starting in February 2020. Externally, it's based on a Rick Dickinson design and looks very similar to a Spectrum 128 – albeit with an integrated SD card slot. Inside sits a speed-boosted Z80 processor, 512kB of RAM and, if you ordered the Accelerated model, a Raspberry Pi Zero accelerator.

Sources

Interviews with Charles Cotton, Rupert Goodwins, David Karlin, John Mathieson, and Nigel Searle, plus email correspondence with Richard Altwasser.

1. Tim Hartnell, Review: Sinclair ZX Spectrum, Your Computer Magazine, June 1982, page 20
    archive.org/details/your-computer-magazine-1982-06/page/n19/mode/2up
2. Nick Hampshire, *Sinclair strikes back at the BBC*, Popular Computing Weekly, 6 May 1982, page 10
    archive.org/details/popular-computing-weekly-1982-05-06/page/n9/mode/2up

3. Brendon Gore, *Interview: The engineer behind the Spectrum*, Your Computer Magazine, July 1982, page 38
    archive.org/details/your-computer-magazine-1982-07/page/n37/mode/2up

4. Steven Vickers, Sinclair ZX Spectrum BASIC Programming manual, chapter 16
    worldofspectrum.org/ZXBasicManual

5. Claudia Cooke, *Modest award-winner sets the pace in micro design*, Sinclair User, August 1982, page 55
    archive.org/details/sinclair-user-magazine-005/page/n53/mode/2up

6. Nick Hampshire, *Sinclair strikes back at the BBC*, Popular Computing Weekly, 6 May 1982, page 10
    archive.org/details/popular-computing-weekly-1982-05-06/page/n9/mode/2up

7. David Tebbutt, Benchtest: Sinclair ZX Spectrum, Personal Computer World, June 1982, page 118

8. Staff reporter, *Spectrum will carry the Sinclair colours*, Practical Computing, June 1982, page 47
    archive.org/details/PracticalComputing1982June06/page/n41/mode/2up

9. Christian Donlan, *Obituary: Rick Dickinson, industrial designer of the ZX Spectrum*, Eurogamer, 29 April 2018
    eurogamer.net/articles/2018-04-29-obituary-rick-dickinson-industrial-designer-of-the-zx-spectrum

10. Floppy Days, Interview with John Grant, episode 85, 8 July 2018
    floppydays.libsyn.com/floppy-days-85-interview-with-john-grant-developer-of-zx80-os-basic

11. Ian Adamson and Richard Kennedy, *Sinclair and the Sunrise Technology*, Penguin Books, 1986, page 127

12. Staff reporter, *ASA criticises Sinclair*, Sinclair User, January 1983, page 14
    archive.org/details/sinclair-user-magazine-010/page/n13/mode/2up8

13. Sinclair Research Ltd advertisement, The Guardian, 8 July 1983, page 7

14. Staff reporter, *Spectrum famine*, Sinclair User, July 1984, page 13
    archive.org/details/sinclair-user-magazine-028/page/n11/mode/2up

15. As above

16. Nick Hampshire, *Sinclair strikes back at the BBC*, Popular Computing Weekly, 6 May 1982, page 10
    archive.org/details/popular-computing-weekly-1982-05-06/page/n9/mode/2up

17. Steve Vines, *Acorn prepares secret 'micro'*, The Observer, 25 April 1982, page 24

18. Sinclair ZX Spectrum advertisement, Sinclair User, August 1982, page 11
    archive.org/details/sinclair-user-magazine-005/page/n9/mode/2up

19. Exatron Stringy Floppy, Wikipedia, version edited 30 April 2020
    en.wikipedia.org/wiki/Exatron_Stringy_Floppy

20. John Gilbert, *Microdrives are still being developed*, Sinclair User, November 1982, page 66
    archive.org/details/sinclair-user-magazine-020/page/n65/mode/2up

21. ZX Interface 2, Wikipedia, version edited 1 November 2017,
    en.wikipedia.org/wiki/ZX_Interface_2

22. Staff reporter, *Sinclair in pub clash with rival*, The Times, 24 December 1984, page 1

23. Rodney Dale, *The Sinclair Story*, Gerald Duckworth & Co Ltd 1985, page 174

24. Chris Owen, ZX Spectrum+, Planet Sinclair
    rk.nvg.ntnu.no/sinclair/computers/zxspectrum/specplus.htm

25. Staff reporter, *£12m Maxwell rescue deal for Sinclair*, The Times, 17 June 1985, page 1

26. Bill Johnstone and Patience Wheatcroft, *Dixons in Sinclair rescue bid as Maxwell pulls out*, 10 August 1985, page 1

27. Maggie Brown, *Sinclair £20m debts force sale to rival*, The Guardian, 8 April 1986, page 1

28. *Alan Sugar, What You See Is What You Get: My Autobiography*, Pan Books 2011, page 298,
    ISBN-13 978-0330520478

Dragon 32

The Welsh dragon that
(briefly) breathed fire

Beige and boxy it may be, but the story of the Dragon 32 is anything but boring. In a narrative arc that fits in perfectly with its mythical name, it burst onto the British computing scene in a blaze of metaphorical fire, only to be slain by two mighty knights known as Sir Clive and Lord Sugar.

From a Welsh point of view, it started off like any good story: in Swansea. Tony Clarke was then financial director of Mettoy, makers of Corgi die-cast models, but in a previous incarnation he'd been an electrical engineer. In 1980, he bought an Apple II, and it got him thinking. 'I started to look at what it did – and to compare it with the other machines on the market,' he told Popular Computing Weekly in late 1982 [1]. 'It struck me that our company could do a better job – in terms of value for money.'

His thinking was straightforward: Mettoy had the experience and tools you needed to assemble a computer, so they wouldn't need to subcontract out manufacturing. It had over 50 years of experience of selling toys to children, with its traditional core market being three-year-olds to 14-year-olds. The keyword there being 'traditional'. 'The over-nines now buy electronic goods – computers, video games, tape recorders, and televisions,' said Clarke.

With Clarke at the helm of the project, Mettoy was also in the fortunate position that it had someone who knew computers inside and out. Quite literally. Together with Gerry Quick, a former Mettoy colleague who handily had a PhD in computer science, he told Popular Computing Weekly that he mapped out the core specification for the machine he wanted.

Clarke now needed to convince the rest of the Mettoy board to back this expensive project. His masterstroke was to hold the relevant board meeting at the 1981 Personal Computer World show. 'They saw hundreds of kids hammering away at keyboards programming micros in ways they couldn't begin to comprehend and they were convinced,' said Clarke. This released funds for him to commission PA Technology, a Cambridge-based consultancy, to build the prototype.

We should now introduce another key player in the Dragon story: the Tandy TRS-80 Color Computer, or CoCo for short. While never a big hit in the UK, it predates the Dragon 32 by two years. Aside from the colour of its chassis (grey rather than beige), it looks almost identical, right down to the keyboard.

This wasn't a coincidence. Ian Thompson-Bell, who took charge of the project for PA Technology, vividly remembers his first meeting with Clarke and Mettoy. 'They came along to the very first meeting with a Tandy CoCo,' he recalls, 'and the data sheet for the Motorola SAM [synchronous address multiplexer] chip. And they said, if you look at this data sheet, you'll see this computer is pretty much a copy of the data sheet. We'd like one like that, please.'

The most important component on that sheet was Motorola's 6809 processor. While still an 8-bit chip, this lifted itself above the Z80 and 6502 by including a 16-bit stack pointer that 'allows arguments to be passed to and from subroutines with ease', to quote Motorola's data sheet. 'That's what turns it from just another 8-bit micro on steroids into a programmer's dream,' says Thompson-Bell. 'It was a real game-changing micro. It was just so much better than the 6502, the Z80, from the point of view of being able to write efficient and effective programs.'

The 6809 may have been a programmer's dream, but the accompanying 6847 video chip certainly was not. 'The colour graphics chip in the CoCo was an NTSC chip designed for 525 lines at 60Hz. The first thing I had to do was come up with some means of getting that chip to do PAL [625 lines at 50Hz].'

For this he needed help from Motorola itself, which is why 'the second meeting we had with Dragon was with the Motorola sales guy,' says Thompson-Bell. He would prove to be a great help when working around the limitations of the video chip. His name? Robin Saxby, now Sir Robin Saxby due to his work as CEO of Arm Holdings.

In hindsight, though, Thompson-Bell thinks Dragon should have used an ASIC – a custom-made integrated circuit, or what Ferranti termed an Uncommitted Logic Array (ULA) – to generate the video. 'We used to have problems with what we call castellations,' he says. 'One of the things with PAL is the lack of synchronism between the colour-burst signal, the video signal itself, and the processor clock. The Commodore 64 overcame that with an ASIC: it had a single big clock generator that did everything and tied everything together. So if there were castellations they were static. On Dragon, because the two clocks were not running together, you could often see the little castellations running up and down on the characters.'

If Mettoy had predicted it would sell 50,000 Dragon 32s then it would have made financial sense to go down this route. As it was, the company had budgeted for sales of 10,000. While it's easy to be critical of this in hindsight, poor predictions were to haunt Dragon Data throughout its brief life.

But that was for the future. For now, Thompson-Bell had more immediate technical problems to solve. One was to replace Tandy's serial port with a parallel port, as this was more compatible with printers on sale in the UK at the time. 'We also used that port to drive a very simple A2D [analogue-to-digital] converter for the joysticks, so we actually had better joysticks than the bang bang ones in the CoCo,' he notes. This meant Dragon owners would benefit from a proportional control: rather than only going up, down, left and right, you could move the stick at 45 degrees and the game would respond.

'It was a really crazy timescale,' says Thompson-Bell. 'I mean very, very crazy. The Dragon 32 never was the Dragon 32: it was the Dragon 16, that's what it was designed as. And two weeks before mass production was due to start, Clive came out with the Spectrum. And Mettoy said, we've got to beat him on every front, so the Dragon is now the Dragon 32. Which was challenging.' This is why, if you open up early Dragon 32s, you'll see a daughterboard that adds another 16kB of memory.

While Thompson-Bell was working on the hardware, Mettoy contracted a Motorola microprocessor expert to work on the BIOS. That expert was Duncan Smeed, then a member of the University of Strathclyde's department of computer science but who would later join Dragon full-time. 'I remember the keyboard routines were actually faster because the PA consultant put a small hardware tweak in it which meant that I could scan the keyboard in a slightly different way, using sort of bit shifting and a bit of twiddling,' said Smeed [2]. 'And because that was done by polling, rather than interrupt, it meant that our BASIC ran faster than the BASIC on Tandy Color Computer.' It also appears to have led to the annoying bug for quicker typists, where tapping out 'the', for instance, often resulted in 'te' appearing instead.

Time to make some prototypes, a burden that fell to Mettoy's Lydon Davies who describes 'hand-soldering the first batch of prototypes (around 20) at home, into the small hours. Unfortunately, due to the time it took to create the dies for the plastic moulding of the case, these prototypes had to have cases hand-cast from dental material, which meant they weighed a tonne, but it did mean examples could be sent out for review etc.' [3]

Davies also confesses that his 'work on the prototypes also involved many trips to the Swansea Tandy (Radio Shack) to purchase TRS-80s, accessories, and software to 'disassemble'. 'This bears out [sic] by the fact that the first game ever played on a

Dragon was TRS-80 *Quest*.' We'll leave that quote floating for any remaining Tandy UK lawyers to contemplate.

Fortunately, Tandy took a pragmatic approach and put the Mettoy board's collective mind at rest by announcing that it wasn't going to sue. 'We have absolutely no objection to the Dragon machine,' said John Sayers, then the managing director of Tandy UK, in the May 1983 issue of Dragon User [4]. 'It is true that the two machines are similar in a lot of respects – they use the same or very similar ROM pack, for example – but I can tell you categorically that we are not planning any legal action.'

One crucial early outing for the prototype was to Boots in May 1982, where it won over Anton Boyes, who was responsible for the company's home computer sales. At that point Boots was still working on its in-store offering of computers, and the Dragon 32 beat off all opposition because it was 'effectively a finished product' [5] while offering lots of memory and high-resolution graphics.

Dragon also took the same decision as Tandy and implemented Microsoft's popular BASIC, helping to lend some extra respectability to the newest computer on the block when it appeared on British shelves in August 1982. That late decision to double the RAM to 32kB distanced it further from the toy-like VIC-20 with its measly 5kB helping, while the typewriter-style keyboard leant it a professional air compared to the still-new Spectrum.

Clarke was also keenly aware that its target audience wanted to go through the traditional route of buying something: £199.50 is roughly £700 in today's money when adjusted for inflation, so hardly an impulse purchase. He believed that people would want to touch it and play with it before spending their cash. This is why all Dragon 32 computers were sold through high-street retailers, typically Boots and Dixons.

Nevertheless, Clarke knew the importance of early computer magazines when it came to the buying decision, and started shipping pre-production versions of the Dragon to reviewers. The first came in Popular Computing Weekly on 8 July 1982, where it received a cautious welcome. The key plus points: its 'smart clean lines', 'strong and efficient' internal design, and the fact its high-resolution graphics were 'so much faster than on the Spectrum'. But it added the caveat that Dragon would have to 'come up with the software cassettes and cartridges it has promised [and] be able to supply and manufacture the computer without suffering the hitches and setbacks which have bedevilled other computer companies.' [6]

There was good news from the manufacturing point of view. Mettoy's established skill of producing mass market goods, and its existing facilities, meant that churning out thousands of machines wasn't an issue. The fact that it was largely using an established reference design for the core electronics also eased the pain. Even the fact it massively underestimated the initial demand didn't cause supply issues: Mettoy had planned on shipping 10,000 computers by the end of 1982, but in the end it produced over 30,000.

Despite this, things weren't running quite so smoothly for Mettoy as a whole. Between 1980 and 1982 it had lost almost £12 million, so chose to hive off its Dragon project and sell it to a consortium in November 1982. This raised a welcome £900,000 for Mettoy, while still keeping a 18.61% stake in the newly formed Dragon Data. There were two other crucial shareholders: Prutech, the high-tech investment arm of Prudential, owned 40.74% and the Welsh Development Agency 22.12%.

Its new independence meant Dragon Data could invest in future plans for growth. And these were ambitious: managing director designate Tony Clarke told Popular Computing Weekly in November 1982 that 'February or March next year should see a disc-operating system and discs... We will be operating a 40 tpi [tracks per inch] drive. The operating system and discs will be available together for around £250.' [7]

Nor did he stop there, with the promise of a 'special version for education – with built-in RGB monitor and cassette player – and a networking system is being developed'. And: 'There will be an expansion box early next year, giving a 64K Dragon. The expansion kit will include the OS9 system, an editor/assembler, and O9 BASIC – all for less than £150.'

Clarke also addressed the reviewer's point about software. 'We have a range of small business software – using the disc system – planned for the spring. This will be followed by more games, home utilities – again, making use of the discs – and a range of educational software for schools.'

There was some software available at launch, but it's hard not to feel sympathy for children who were given a Dragon 32 for Christmas in 1982. While their ZX81 or ZX Spectrum friends had shelves and shelves of games to choose from, Dragon 32 owners had to pick from a handful, many of which were bizarrely packed in polythene bags.

They had a choice of six ROM cartridges released by Dragon, which had the advantage of loading instantly and the disadvantage of costing £19.95 a pop.

Highlights included *Meteoroids*, an *Asteroids* knock-off, *Starship Chameleon* (where you destroyed enemy rockets by colliding your spaceship into them), and the high-octane, robot-zapping *Berserk*. Dragon also provided a bunch of games on cassette, with prices a more reasonable £7.95.

Third-party games and software did start to appear, thanks in large part to John Symes. He was the man behind Cornwall-based Microdeal, which made most of its money by adapting games developed for the Tandy CoCo. However, he didn't get much assistance from Dragon. 'Dragon has been of no help whatsoever to any of the software houses; they didn't even tell us they had reconfigured the RAM,' he said [8]. 'It meant we had to withdraw two games.'

As was common for small British software houses of the early 1980s, Microdeal also skirted the line when it came to copyright and trademarks. One of its biggest hits was *Donkey King*, which was essentially a copy of Nintendo's *Donkey Kong*. Following a stiff letter from Nintendo's lawyers, Microdeal renamed it *The King* in the summer of 1983.

Richard Harding, who runs the retro and archiving website **dragondata.co.uk**, doesn't think Dragon owners fared too badly. 'There was a small set of games, but most of them were pretty good,' he says. 'There were some really good adventures and I used to play *Lunar Rover Patrol* a lot. I never thought I was wanting for games.' And, besides, he explains, if he wanted to play *Deathchase*, *Manic Miner*, or *Ant Attack*, he could always nip round to a friend's house.

In May 1983 the Dragon even got its own magazine in Dragon User. It started modestly with 52 pages, with most of its advertising dominated by software packages available by mail order – how could you not be tempted by Ohm, 'a program allowing you to simulate an experiment to verify Ohm's Law and test understanding with problems' [9]? Yours for £9, including postage and packaging. The magazine is still fondly remembered by Dragon owners for its mix of news, interviews, software reviews, and BASIC listings.

Tony Clarke appeared in his second interview for the magazine in July 1983, at roughly the same time as the Dragon 64 was released. He was as full of enthusiasm and optimism as ever, extolling the virtues of the new £275 disk drive system and the OS9 UNIX-like operating system. For professional users, there were obvious merits to OS9: a high-quality spreadsheet and database, C and Pascal programming languages, a dedicated assembler. 'It gave you high-res text as well,' says Harding. And for the normal, everyday user? 'They just wouldn't need it at all.'

Clarke did need to row back on his earlier promise that Dragon 32 owners would be able to upgrade to a Dragon 64 via an 'expansion box' that would include 'the OS9 system, an editor/assembler, and O9 BASIC' for less than £150 [10]. In the event, Dragon 32 owners were offered the chance to upgrade to a Dragon 64 by returning their old Dragon 32s and including a cheque for £140. But this didn't include OS9, which cost another £50.

Even before the Dragon 64 went on sale, though, Dragon Data went through the first of its financial crises. A slow summer meant less cash coming in while expenses were rising: quite aside from all the development costs associated with the 64, it was also hard at work on the new home computer and small business machines that Clarke had alluded to. It needed cash, and it got it via a £2.5 million injection from its shareholders.

Dragon attempted to spin this as an investment into a growing company, but when news broke that Clarke had 'stepped down' while a new MD was being appointed from electronics giant GEC, people used the 6809's hardware multiplication instructions to see what happened when two and two came together. After all, Prutech was a major shareholder in both Dragon Data and GEC. Could this proviso have been part of the £2.5 million package?

There were signs that the pressure of running a rapidly growing company, with so many shareholders with different priorities, were affecting Clarke. 'He began to keep changing his mind,' says Thompson-Bell, who has a huge amount of sympathy for the pressure Clarke was under. 'We would agree something in one meeting and say, fine, we'll go away and do that. And then we come back and say, OK we've done that. He'd say, no, I didn't want that, I wanted this.'

Whatever the reasons for Clarke's departure, Dragon User featured a two-page interview with the new man in the Dragon Data hot seat in its December 1983 edition. Unlike Clarke, Brian Moore (not, sadly, the football commentator) wasn't too interested in the technical side of the operation, even if he had previously been deputy MD of a 'GEC subsidiary specialising in microprocessor-controlled heating and ventilation systems' [11]. He wasn't even sure how long he would be in charge, as officially he was on secondment from GEC.

By this time, the Dragon 64 had gone on sale, but in the US rather than the UK. This happened through a joint venture with Tano Corporation, and to understand the lack of a splash it made consider that it merited a mere three-line news story in the influential Byte magazine [12]: 'The British-designed Dragon computer is being manufactured and sold in the U.S. by Tano Corp. (New Orleans, LA). For $399, the machine includes 64K bytes of RAM, sound, 256 by 192 color graphics, and Extended Microsoft BASIC.'

It wasn't received with wild enthusiasm by the UK computer press either. Aside from the addition of an RS-232 port, and the 64kB of memory, it was essentially identical to the Dragon 32, which meant there was no obvious point to upgrading (you could already buy RS-232 adapters and memory expansion packs for the Dragon 32). It did fix the typing bug mentioned earlier, and added an extra DIN port, but Dragon 32 user David Stobie couldn't get excited by the 64 – even its fancy new grey livery.

For the 'ordinary man in the street there seem to be few advantages and even some disadvantages,' he wrote [13]. 'The 64 is really a 32 with extra facilities stitched on instead of a really new machine. Execs and Peeks and Pokes are needed to use most of the new facilities when they should be an integral part of the machine.'

Was there any benefit of having 64kB of RAM? The answer was that only small businesses need apply. This would allow them to use OS9 and the advanced software packages available for it, but even then they had to wait until spring 1984 – and buy a Dragon DOS disk system. Technically, OS9 was an excellent operating system, but it's notable that even Dragon Data's advertising for the Dragon 64 steered clear of mentioning it.

With sales struggling over Christmas 1983 and the spring of 1984, Dragon Data turned to GEC McMichael – one of the many GEC subsidiaries – for a lift. In an interview with Dragon User, GEC McMichael's Ron Bosanko was clearly upbeat about Dragon's prospects. 'Ron finds his enthusiasm hard to conceal,' wrote Graham Cunningham [14], 'talking of the pivotal role micros can play at one end and the TV screen at the other. But he refuses to be drawn on this for the moment, promising only that ideas being discussed now should yield some interesting results by the end of the year – "moving things a little bit beyond the field of home computing".'

Unfortunately, Dragon didn't have that long. In June 1984, at precisely the same time that Dragon User readers were enjoying Bosanko's optimistic predictions for the future, Dragon Data called in the receivers. It needed to find a new buyer, and over the course of the summer it variously looked like Tandy and GEC would pick up the reins. In the end, though, a mysterious, newly formed Spanish company called Eurohard SA stepped in.

Time for yet more optimism. 'I hope this is the beginning of a whole new Dragon era,' said Justo Alverez [15], Eurohard's director for industrial engineering. His plan: to move construction of all new Dragon 32 and 64 computers to Spain, to build a Microsoft MSX machine, and to work with Spanish broadcasters to create a similar programme as the BBC. In effect, for the Dragon to become the Spanish BBC Micro.

Lydon Davies was one of the few Dragon Data employees to make the move to Eurohard. 'After the sad downfall of Dragon Data I stayed on and moved to Spain with Eurohard and helped them set up the initial production,' he said [16]. 'This was in 1985 and the factory was located in Cáceres in the Extremadura part of Spain. It was a very poor region and I always remember my first day as the bus I arrived in drove over the main power cable to the factory – after this I always stopped the bus early and allowed it to bump over the cable without me on board. It was quite amazing how they purchased Dragon Data and ever made anything as they didn't appear to have any real money either.'

According to an interview with the Eurohard president Eduardo Merigo in the January 1985 edition of Dragon User, production started in November 1984 with a plan to create 'enhanced' versions of the Dragon 32 and 64 – called the Dragon 100 and Dragon 200 respectively – in March 1985. While the Dragon 200 was designed, and has even been exhibited at the Design Museum of Barcelona, none were ever sold. Likewise, only one episode of the suspiciously expensive TV programme aired.

It was a suitably bizarre end to the story of a computer that burned so bright. Then puff. It was gone.

Don't let Dragons go extinct

While there are big retro movements supporting the likes of the Amiga, C64, and Spectrum, it's harder for the computers that never sold hundreds of thousands of units. There remains a lot of affection for the Dragon 32 and 64 among its owners – and lapsed owners – but with fewer of them there's a greater chance of games and information being lost.

This spurred both Richard Harding and Simon Hardy to set up Dragon archives to capture everything related to the released computers and those that didn't make it beyond the prototype stage (of which there are surprisingly many). Harding runs **dragondata.co.uk**, Hardy the wiki-style **archive.worldofdragon.org**.

If you're a lapsed Dragon user, you may well be able to help. "Some software hasn't been archived yet," says Harding. "It does upset me because there are a lot of people who collect things and keep it to themselves. I want to keep the Dragon alive."

So, if you have a Dragon in the attic, along with a bunch of games, then take a look at their sites and see if you can fill the gaps. Don't let Dragons go extinct.

Sources

Interviews with Richard Harding and Ian Thompson-Bell.

1. David Kelly, *The number of the beast*, Popular Computing Weekly, 18 November 1982, page 10
   archive.org/details/popular-computing-weekly-1982-11-18/page/n9/mode/2up
2. Jason Fitzpatrick, The Prototype Dragon 32 and Dragon Alpha with Duncan Smeed, Centre for Computing History, 6 May 2019,
   youtu.be/0IipxReq0G8
3. Lydon Davies, Dragon Data The Archive,
   dragondata.co.uk/history/Quotes/index.html
4. Staff reporter, *Tandy UK puts Dragon at ease*, Dragon User, May 1983, page 9
5. Graham Cunningham, *Boots provides a firm footing for Dragon's future*, Dragon User, June 1983, page 20
6. PCW staff, *Enter the Dragon*, Popular Computer Weekly, 8 July 1982, page 12
   archive.org/details/popular-computing-weekly-1982-07-08/page/n11/mode/2up
7. David Kelly, *The number of the beast*, Popular Computing Weekly, 18 November 1982, page 10
   archive.org/details/popular-computing-weekly-1982-11-18/page/n9/mode/2up
8. Graham Taylor, *Microdeal fills software gap*, Dragon User, May 1983, page 23
9. Advertisement, Learn Physics with your Dragon, Dragon User, May 1983, page 14
10. David Kelly, *The number of the beast*, Popular Computing Weekly, 18 November 1982, page 10
    archive.org/details/popular-computing-weekly-1982-11-18/page/n9/mode/2up
11. Graham Cunningham, *New man in the driver's seat*, Dragon User, December 1983, page 34
    archive.org/details/dragon-user-magazine-08/page/n33/mode/2up
12. Microbytes, Byte Magazine, September 1983, page 8
13. David Stobie, Dragon 64, Your Computer, January 1984, page 70
    archive.org/details/your-computer-magazine-1984-01/page/n69/mode/2up
14. Graham Cunningham, *Dragon finds a new lair*, Dragon User, May 1984, page 30
    archive.org/details/dragon-user-magazine-13/page/n29/mode/2up
15. Gordon Ross, *The Spanish connection*, Dragon User, November 1984, page 10
    archive.org/details/dragon-user-magazine-19/page/n9/mode/2up
16. Lydon Davies, Dragon Data The Archive, **dragondata.co.uk/history/Quotes/index.html**

Commodore 64

The world's biggest
selling computer...
until 2019

The Commodore 64 is arguably the most important computer ever built. It certainly sold more than any 1980s rival, with Guinness World Records stating that around 30 million C64 computers shipped during its staggering twelve-year lifespan*. Compare that to the Apple II series, often cited as the world's most influential computer, which sold around six million.

While Commodore executives were confident that its computer would be a big seller, even they were surprised by the scale of its success. 'When we launched it in the States,' says Kit Spencer, who was in worldwide charge of marketing for Commodore at the time, 'I held a press conference. There hadn't been a million computers sold worldwide at the time, and I wanted a headline, so I said we're going to sell a million computers in the next year... I went back a year later for another press conference and said, I'm sorry, I lied to you last year. I said we'd sell a million computers. We didn't: we sold one and a quarter million.'

To understand why Spencer's initial statement was so bullish, we need to sink ourselves back into the world of computing in 1982. Thanks to Clive Sinclair's efforts to bring low-price computers to the market, British homes had an unusually large number of micros in them. The rest of the world wasn't so fortunate, with American homes still dominated by cartridge-based gaming systems made by Atari.

Commodore already had form when it came to shaking up the market. Its VIC-20 may 'only' have sold a million units during its brief life, from 1981 to 1984, and most of those sales came late on, when Commodore discounted it to below $100, but it had shown its founder and driving force Jack Tramiel that there was a mass market for computers. Especially computers that could play games.

Never one to rest on his laurels, Tramiel was always looking for the next big thing. While other company leaders may have wanted to protect sales of their big seller – and make no mistake, the VIC-20 was a huge seller by the standards of 1981 – it took Tramiel a matter of minutes to be convinced that Commodore should develop its bigger, better successor.

And it took just two people to persuade him: Charles Winterble and Al Charpentier. Both worked for MOS Technology, which Tramiel had snapped up in the mid-1970s

---

* According to David John Pleasance, managing director of Commodore UK between 1993 and 1995, the actual figure was a tad under 27 million. 'I can tell you that because, when we were thinking about doing a management buyout, we got access to all the figures.' [1]

so that he didn't need to rely on Texas Instruments for calculator chips. It proved to be one of the wisest pieces of business in the history of computing, as he also inherited the services of Chuck Peddle – the architect of the hugely influential 6502 processor, which would go on to power the Commodore PET, Apple II, and VIC-20.

By 1981 Chuck Peddle had left to create his own business, and Winterble was in charge of engineering. He and Charpentier were keen to produce a successor to the VIC (video interface chip) that both powered the graphics and provided the name of the VIC-20, and the top item on their agenda was its 22-column limit. This meant it could only show 22 characters on one line, which made the graphics look toy-like when compared to more sophisticated computers such as the Apple II. They wanted to create a 40-column version that could be used in gaming systems but also power a next-generation computer.

Tramiel needed little persuading. In Brian Bagnall's book, *Commodore: A Company On The Edge*, Winterble describes the first time he raised the idea with the cigar-smoking boss. 'We want to do a 40-character game chip and it can also be a computer,' Winterble said [2]. 'We want to organise a project to do it.' According to Winterble, Tramiel's response was instant: 'I didn't even really finish talking and he said, "Do it."'

Rather than rush in, Winterble gathered the brainiest of the MOS Technology brains to literally dissect the market – they bought and disassembled all the available products – and work out, from scratch, what they needed to create. The key was to find out what could be done now and look into the future to see what features their chip would need to include to compete.

One feature was crucial: sprites. Think back to the games you played on the Commodore 64; they involved moving graphical elements around the screen. Rather than needing the computer to create those effects in software, which involved a painful redraw process, sprites allow the graphics chip to put the image on screen in hardware. All the programmer needs to do is tell it where on the screen the image should go and where it should move to next.

Sprites weren't new. Via an Extended BASIC cartridge, the Texas Instruments TI-99/4A featured a single-colour sprite that did much of what the MOS Technology team desired. Atari also included four sprites in its computers. Naturally, however, Charpentier and his team wanted more: more types of sprites, more colours. They added two more features that would prove useful to software developers too: collision detection and the ability for sprites to double in size.

Meanwhile, Robert Yannes – the young engineer so crucial in the VIC-20's creation – had his mind set on audio. In particular, synthesizers. Having purchased his own synthesizer kit back in college, he now wanted to make this cutting-edge technology available to mainstream computer users. He had already made some audio contributions to the VIC-20, but the Commodore 64 went several steps further: it would have its own sound chip.

Winterble set some parameters for Yannes, including cost and die size. He worked with the 24-year-old to create a list of 'must-haves and might-dos'. The work itself, however, was Yannes's own. In Bagnall's book, he explained the parlous state of audio in computers until the Commodore 64 came along. 'With most of the sound effects in games, there is either full volume or no volume at all,' he said[3]. 'That really makes music impossible. There's no way to simulate the sound of any instrument even vaguely with that kind of envelope, except maybe an organ.' Yannes's chip would eventually include three 'voices' and was capable of producing four base sounds. A head-and-shoulders improvement over the existing competition.

Nine months after the initial meeting with Tramiel, Commodore had two working chips: the VIC-II and Yannes's SID (Sound Interface Device). Even then, though, the Commodore 64 concept hadn't taken shape. MOS Technology had created these chips for anyone to buy, and in effect they were just two ingredients – albeit crucial ingredients – in the recipe to make a new computer or games console.

There was another hurdle too: the VIC-20 was selling almost too well. While Jack Tramiel may have been convinced of the need for a next-generation computer, other members of Commodore senior management were resistant to the idea of confusing the market by introducing a second consumer-focused computer, and potentially killing their golden goose in the process.

Plus Commodore was busy on other projects. Yashi Terakura, one of the two Japanese engineers who was so pivotal in the final design and delivery of the VIC-20, was already using the VIC-II and SID in the Commodore Max Machine, which was primarily designed for the Japanese market. The MOS Technology engineers were also meant to be working on the P series and B series: personal and business successors to the PET. All three new computers launched in 1982, the same year as the Commodore 64.

Against this background, it's perhaps surprising that Jack Tramiel's determination to press ahead with the 'VIC-40' – the Commodore 64's codename – never wavered. If anything, he wanted to double down. Where previously he had leaned towards less memory in computers to keep costs low, this time he wanted the new computer to have 64kB of RAM. His reasoning was simple: the new Apple II included 48kB, as did the Spectrum 48K, so the VIC-40 would immediately stand out from the crowd. And he believed that 64kB RAM chips, which were just starting to be produced, would have dropped significantly in price by the time the VIC-40 hit the shelves.

Tramiel gave the go-ahead for the project in November 1981, and this time the surprise he had for his engineers was less welcome. Rather than target the Consumer Electronics Show (CES) scheduled for summer 1982, seven months away, he wanted to demonstrate the prototype at the January 1982 CES. Six weeks away. Winterble handed this daunting task to Robert Yannes, who was already familiar with the VIC-II graphics chip and, of course, the SID sound chip he had created.

Yannes wasn't a systems engineer, so Robert Russell (another crucial figure behind the success of the VIC-20) stepped in to help design the hardware architecture. Yannes kept this deliberately minimalist – you won't see much if you break open a Commodore 64 – and also decided not to base it on the VIC-20's architecture. There was good news for VIC-20 upgraders, though, with the same serial, cassette, and user port in place. And you didn't need to throw away your joystick either, with two ports available.

The cartridge design was borrowed from Terakura's Max Machine, with Yannes solving the rest of the technical puzzle via a programmable logic array (PLA) chip. This also bought him some time: 'I didn't have time to design all the logic before they laid the PC board out,' said Yannes [4], 'so I just took a PLA and named the signals I needed and told them to lay that out.'

With January rapidly approaching, the MOS Technology team took the pragmatic decision to use the same case and keyboard design as the VIC-20, right down to the four programmable keys. 'We just put it in a VIC-20 case and spray-painted it,' Yannes said. 'Everything about the Commodore 64 is the way it is because of just an unbelievably tight time constraint on the product.'

Up until CES, the VIC-40 project was still a secret. Not even senior Commodorians such as Kit Spencer were told about it. It wouldn't stay that way for long: Jack Tramiel used CES to announce to the world that the new computer (still unnamed) would sell for $595. It was an astonishing price considering its graphical and audio capabilities, never mind the promised 64kB of memory – the 16kB Radio Shack TRS-80 cost $599, while the newly released IBM Personal Computer cost $1,265 with 16kB of RAM.

Perhaps most importantly for Tramiel, however, the Commodore 64 would hugely undercut the Apple II Plus with its 48kB of memory. While the exact price you paid would depend upon your reseller, it would be well into four figures. By this time, Commodore had already earned a reputation for making big claims about new products and then quietly canning them, leading its competitors to suggest that this was another such boast. Plus, that price seemed way too good to be true.

Kit Spencer would later use the cynicism of its competition against it in a full-page ad stating: 'When we announced the Commodore 64 for $595, our competitors said we couldn't do it. That's because they couldn't do it.' [5]

While the launch may have been met with cynicism from Commodore's competitors and even the press – the announcement gained little coverage in the magazines of the time – it received a much warmer reaction from show attendees. 'The C64 just kind of blew everybody out of the water because it came out of nowhere,' Yannes said [6]. 'There was no expectation of it, it was very reasonably priced, and it had 64K of RAM, which was a magic number at that point in time because nobody else had 64K of RAM.'

It helped that, by the standards of the time, the prototype demoed at CES was surprisingly complete. The MOS Technology team had even managed to get it working with its own printers and disk drives. The positive feedback and seeming readiness made Jack Tramiel all the more eager to produce machines quickly, and he set a deadline of three months to move from prototype to finished product.

This proved too ambitious. While Charles Winterble and his team were talented engineers, they had battles across several fronts. First, they were still fighting bugs in the two new chips, the VIC-II and SID, that were so pivotal to the C64's eventual success. Second, they were hindered by the decision to use the same casing as the VIC-20, which was really too small to house the new components. And third, Commodore's engineering resources were spread thin: Robert Russell estimated that, at that time, there were only around 15 engineers capable of the design work needed across software, hardware, and silicon. And several of them were already committed to other projects.

There were more internal challenges too. Due to the secrecy surrounding the VIC-40 project, no one in Commodore's marketing team knew about the computer until just before CES, and that meant that some crucial decisions had been made without consulting people who knew the market. In particular, Kit Spencer had seen the rise of the CP/M operating system and felt that any serious computer released in 1982 – particularly one that hoped to sell to business users – should support it, and thus all the business software that ran on it. (The eventual solution was to sell a CP/M cartridge complete with its own Zilog Z80 processor, but it suffered from some serious problems and never sold in bulk.)

Spencer was also an advocate for software compatibility. 'That's always a bit of a Catch 22 because you do want compatibility but you want progress as well. And when you make that quantum leap to something else, and you lose compatibility, it's always a bit of a problem. So I would be arguing with the engineers, if I was arguing at all, to try and keep some compatibility on machines as you went on.'

It was an argument he would ultimately lose, with Tramiel – no doubt aware of setting back the launch date – backing the engineers. But there was one crucial argument that Spencer did win. 'I had quite heated discussions with Charlie [Winterble] for a while,' Spencer says, 'because I knew that we'd messed up to some extent the launch of earlier products by having no manuals, no software, no decent packaging, when it came out. And I knew this was a massive product.'

To ensure the launch was successful, then, Spencer had to know all the details of the computer well before launch. 'Eventually I came to an agreement with Charlie. I said, I'm going to hire one guy, a techie guy; you can tell me if he's OK. And he needs 100% access to your team. But nobody else anywhere in the world is going to contact your team for anything.'

As this indicates, Spencer was determined that the Commodore 64 would be the slickest launch in the company's history. Not only would it be sold with a good user manual, packed with BASIC examples, but it would benefit from the Commodore's growing list of consumer dealerships.

Spencer was also responsible for naming the computer. 'Its single biggest feature was 64K of memory, which was more than the Apple's, one of our biggest competitors at the time, or the IBM that had just come out. So I said, let's put the biggest feature in the title. And let's make our company the brand name rather than the VIC or the

PET – that doesn't have the continuity of the company. And it also sounded good: Commodore 64 sort of had a rhythm to it.'

What Kit Spencer also knew was that it needed to launch with a truckload of high-quality software. 'I said, what's our wish list for software at launch? You want a spreadsheet, you want a word processor, you want some games, you want some educational software.' He asked around all the Commodore countries checking who could produce what. 'The UK did quite a lot of software. The spreadsheet, word processors all came from the UK. The UK was quite a hotbed of development – a lot of other stuff, including games, came out of the UK.'

With a limited number of prototype machines to hand around, Spencer could only spare one for the UK, with one apiece shipping to Germany, Japan (for creating games), and to Waterloo University in Canada. 'There was a lot of educational software on the Commodore PET and one of the prototypes went up to Professor Graham at Waterloo University,' says Spencer. 'His students worked 24-hour shifts, I think, eight hours each, converting all the education software on the Commodore PET over to the Commodore 64 on this one prototype!'

This hard work and focus would prove beneficial to Winterble's team, too, as Spencer reported bugs and quirks direct to his contact, who would then pass this on. All this would eventually mean that, when released, the C64 would be the most bug-free computer Commodore had brought to market. And arguably the most bug-free home computer to that point.

There was one big mistake that would come to waste hours of C64 owners' lives; drive speed. Frustrated by the slow read/write speeds of the original disk drive, Robert Russell had added extra high-speed lines to the circuit design. The theory was that this would have removed the speed limits of the serial bus, making the drive 20 to 30 times faster than the VIC-20's. But when MOS Technology sent the design to the west coast engineers, who had the tools to turn them into schematics and final production drawings, those engineers removed the lines – they were under pressure to fit everything into the limited space of the VIC-20's case, and didn't recognise the lines' importance.

Unfortunately, by the time Russell spotted the problem, thousands of boards had been produced, and stopping it in mid-track would have delayed the C64's launch by several weeks and cost the company both lost revenue and lost reputation. So

Commodore took the decision to keep on producing them and abandoned work on a faster drive that would have taken advantage of the high-speed bus.

This problem was compounded by a desire for backwards compatibility with the existing drive: Commodore didn't want thousands of outdated disk drives to be left sitting on shelves. If you ever sat in front of a C64 wondering why it took two minutes to load a program, the many compromises Russell and his team had to make to ensure compatibility, and hit Tramiel's ferocious deadlines, are directly to blame.

Tramiel's ultimatum to produce the C64 quickly proved costly to Commodore too: it had to write off a million dollars' worth of ROM chips because the factory had rushed them into production before Russell had given his final sign-off. The VIC-II chips also faced yield problems and suffered from a glitch where light blue sparkles appeared on a dark blue background once the chip became hot (the so-called 'sparkle bug'). Commodore elected to try to fix this in software rather than hardware.

There were further challenges during production. With little room for error due to the tight confines of the case, workers on the assembly line accidentally screwed through trace lines on the printed circuit board. Early units had intense colours because they were being hand-tuned by workers. Shortages of the VIC-II chips, still suffering from low yields, meant that some production lines were using defective chips simply to get Commodore 64s out into the shops before Christmas.

Despite all these challenges, the Commodore 64 proved a huge success. Together with heavy discounts on the VIC-20, it pushed the company's sales through the roof: in 1978, its annual turnover was $50 million [7]; by the end of 1982, it was reporting sales of $304 million [8].

'The thing that gets lost is that level of organisation that you expect to see in a large corporation,' says Neil Harris, one of the 'VIC Commandos' that had helped to launch the VIC-20 against resistance within Commodore. 'You can only do that when you have a pretty stable environment. And we were pretty much the opposite of a stable environment.'

This meant that communication was less ad-hoc and more entirely random. 'You'd go out to lunch, go play games in an arcade or whatever, and you find out a lot more than you would sitting at your desk,' says Harris. It didn't help that, throughout 1981, he was spending much of his time on the road to help build the network of VIC-20 dealerships, so he was one of the many who knew nothing about the VIC-40 project. 'Somewhere during that year, somebody brought me a 64 and showed it to me and showed me the new user manual. I said, "Why don't you guys let me write this? It's not as good as the one I wrote [for the VIC-20]."'

One of the reasons why Harris was keen to do so – other than being thoroughly bored of a life on the road – was that he could immediately see that the Commodore 64 was going to be a huge success. 'I mean, if you thought the VIC-20 was going to be huge, you have to recognise that the C64 was going to blow the market wide open,' he says.

Early outings of the machine received a similar response from shareholders. 'We got standing ovations from some,' recalls Harris, who was frequently rolled out at shareholder meetings due to his ability to both demo the machine's skills and talk to an audience. 'I'm talking away and I'm playing a game and I'm beating the hell out of the machine. And people just got very excited by that. It was an exciting time. People were getting rich, the sales guys were getting rich, and the shareholders were getting rich. I didn't get rich. But I had a good time anyway.'

Meanwhile, Spencer was keenly aware that Commodore users wanted information about the computers they were buying, whether that was explainers on how to make something work, programming guides, or reviews of upcoming software and peripherals. Indeed, he had created what is almost certainly the first vendor-owned computer 'magazine' (it was more a newsletter in truth) when the PET launched in the UK.

Knowing that Jack Tramiel hated anything that cost his business money, Spencer had charged for that newsletter from the off, so it's little surprise that he gave a warm welcome to Harris when he suggested that he took over the American Commodore magazines. At that time, Harris describes the magazine as 'basically a PR thing talking about how wonderful the company is and with stories with smiling, happy people using Commodore computers – but it wasn't anything that would be worth your money as a subscriber.'

Despite Harris's lack of experience – while he'd written a handful of magazine articles and been a key contributor to the programmers' reference guides, that's hardly a career in publishing – Spencer decided to put him in charge. 'A year later, we had two magazines with a circulation above 100,000 and instead of losing half a million dollars a year, we're making a million dollars a year in profit,' says Harris.

In the wider context of the C64's success, however, this contribution was tiny. Incredibly, Commodore would more than double its revenue from $304 million in 1982 to $681 million in 1983, and the vast majority of that growth came from the C64: 'Microcomputer systems accounted for 96% of overall Commodore sales,' stated its 1983 annual report to shareholders, 'compared to 75% in fiscal 1982 and 71% in fiscal 1981.' [9]

By the middle of 1983, Commodore was finally manufacturing enough C64 systems that it could expand sales to other countries. As ever, the UK was one of the most important. Mike Curtis reviewed the Commodore 64 in the May 1983 issue of PCW, but his review wasn't glowing. 'The Commodore 64 is what you might expect from a major manufacturer like Commodore: a professional, high-quality machine with a guaranteed large software base. There is nothing startlingly new about this machine, in some ways it is a marketing ploy like the new Apple IIe: upgrading a well-tried and proven architecture with the most modern technology.' [10]

The big challenge Curtis flagged, though, was the price. Commodore initially priced the C64 at £344.95, twice the price of the £175 Spectrum 48K – and Sir Clive quickly slashed that to £129.95. This meant the C64's natural competition at the time was what Curtis called 'other more "serious" machines, with an educational and small business market in mind'.

Dick Pountain, who had been writing for the influential Byte magazine and PCW since the days of calculators, was also given a Commodore 64 and was much more impressed. 'It was a neat package, particularly compared to everything else that was around. It was better built. It looked good. And it worked.'

Pountain also had one particular reason to love the C64: the Forth programming language. 'Somebody wrote a Forth for it and I used the C64 to write graphics programs. Which ran like shit off a shovel. Fast. In stark contrast to everything else.'

It turned out that British buyers were more in tune with Dick Pountain's view, even if they preferred to play games using the sprite-powered C64 rather than create their own graphics. By the end of 1983, it was a straight shoot-out between the ZX Spectrum and the more expensive Commodore 64; although Acorn offered the promise of the Electron in the middle ground, for £199, it couldn't deliver systems in time for Christmas.

A year later, in time for Christmas 1984, Commodore dropped the C64's price to under £200, bringing it into reach for most British households, and it would inspire tens of thousands of British schoolchildren to code. While the same could be said of the BBC Micro – and to a lesser extent the ZX Spectrum – Kit Spencer emphasises that the C64's power was its worldwide influence. For him, this puts it as one of the three most significant computers ever produced.

'The first one to really start opening up what I call a reasonable-size market for microcomputers was the Commodore PET,' he says. 'I think the IBM PC coming out really opened up the mass business market and the Commodore 64 opened up the absolute mass market for everybody, for the home market. [The BBC Micro] was very significant in the UK because it was in a lot of schools. It really did get an awful lot of people into computing. But in worldwide terms, and in terms of innovation, I don't think it really opened up a new market.

'You could go to the US and someone will say, what was the BBC Micro? They'd never heard of it. Whereas you could go almost anywhere in the world with a Commodore 64 and they know it.'

The ousting of an icon

In 1965, typewriter manufacturer Commodore was in trouble. It had become embroiled in a financial scandal thanks to its close connections with the Alliance Acceptance Corporation – which shared directors with Commodore – and its financing deals. While Jack Tramiel wasn't indicted as part of the public inquiry, he was forced to sell a big portion of Commodore to Canadian financier Irving Gould.

So started a fraught 18-year business relationship, with Gould controlling Commodore's purse strings while Tramiel did what he did best: sniffed out a market and then set out to dominate it. For almost two decades, the partnership worked well.

With Commodore's explosive growth in the early 1980s, however, cracks started to appear. Tramiel often wanted investment that Gould was unwilling to provide, such as creating its own computer manufacturing facility. In particular, Tramiel felt Commodore should take advantage of its ever-rising share price to raise money for investment and to pay off debts. Gould said no.

There were also personal issues: Tramiel was irked by Gould using Commodore assets, and the company's private jet in particular, for personal trips. Another reported point of conflict? That Tramiel wanted to promote his son Sam into a senior position at Commodore, while Gould was keen to bring in high-profile names from outside. In late 1982, for instance, Gould hired Bob Lane to be president, despite his background being in telecoms rather than computers. Lane didn't last long, but Kit Spencer believes it 'was probably the start of Jack and Irving's long-term relationship going wrong'.

Pressure between the pair also stemmed from the sheer growth of the business. Once it passed a billion dollars, it moved into a different league and Gould apparently thought that Tramiel was no longer the right person to be its figurehead. (A private man, Gould never put forward his side of the argument, and died in 2002.)

At CES, in January 1984, matters reached a climax. 'Jack Tramiel got up on the stage and there's a room with sort of a thousand people in it,' says Neil Harris. 'Jack announced that Commodore had crossed a billion dollars in revenue that year. And I'm in the back of the room looking at him saying, "Why does he not look happy?" And a matter of days later, we found that Jack had decided to leave Commodore. Jack and the board or the chairman had had a falling out and it was all over.'

We will never know for sure what happened between Gould and Tramiel, but his son Leonard – who had made valuable contributions to the development of the PET despite not being a Commodore employee – later heard his father's side of the story. 'He had a very strict moral compass. And if he saw that you weren't doing things the way you should, from a moral point of view, he'd fire your ass. He couldn't fire Irving. So he quit. As my father put the conversation, he said, "As long as I'm President of the company, you can't do this [use company assets for private reasons]." And Irving said goodbye. And dad left.'

To find out what happened next – and the world did not have to wait for long – read the stories of the Commodore Amiga and the Atari ST.

What came next

Commodore Plus/4

Release 1984   **Price** £299

The Plus/4's big selling point was its four built-in software packages – graphics, spreadsheet, database, and word processor – and a much-improved version of Commodore BASIC. People were less enthused by the lack of sprites and its non-standard joystick and cassette interfaces.

Commodore 16/116

Release 1984   **Price** £140

This low-cost version of the C64 sacrificed memory (16kB, as the name implies, rather than 64kB) and with it compatibility: games that required more memory to run simply wouldn't work. While its graphics were also poorer than the C64, it actually ran faster thanks to a 2MHz TED processor.

Commodore 128

Release 1985   **Price** £270

It's too simplistic to call this the Commodore 64 updated with 128kB of RAM: the company claimed it was 'really three computers in one' [11] as it 'can run 64K, 128K and CP/M software'. The latter was thanks to an integrated CP/M module, complete with its own Z80 processor, but the C128 also benefited from support for 80-column displays.

THEC64

Release 2019   **Price** £110

UK-based Retro Games Ltd brought the C64 back to life in 2019 with this full-size replica, complete with fully working keys, a joystick, and a built-in bundle of 64 games. To relive your C64 (and VIC-20) days, all you need is a TV with an HDMI input.

Sources

Interviews with Neil Harris, Dick Pountain, Kit Spencer, and Leonard Tramiel.

1. Martin Cooper, The rise and fall of Commodore, 2 July 2019
    bcs.org/content-hub/the-rise-and-fall-of-commodore
2. Brian Bagnall, *Commodore: A Company On The Edge*, Kindle edition, location 6319
3. As above, location 7385
4. As above, location 7602
5. Commodore advertisement, Byte magazine, 1982, Volume 07, Number 10, October 1982, page 51
    archive.org/details/BYTE_Vol_07-10_1982-10_Computers_in_Business/page/n51
6. Brian Bagnall, *Commodore: A Company On The Edge*, Kindle edition, location 7791
7. Commodore International Limited, Annual Report, 1978, page 3
8. Commodore International Limited, Annual Report, 1983, page 2
9. As above
10. Mike Curtis, *Benchtest: Commodore 64*, Personal Computer World, May 1983, page 136
11. Commodore 128 advert, Practical Computing, January 1986, page 9
    archive.org/details/PracticalComputing1986January01/page/n7

Acorn Electron

The Spectrum killer that almost killed Acorn

In a different universe, the Acorn Electron is the best-selling home computer of all time. This compact, cheap version of the expensive BBC Micro blitzed the opposition thanks to advanced graphical capabilities, a booming market of parents wanting little Johnny (and, as we shall see, Jane too) to develop their programming skills, and a groundswell of support from hardware add-on providers and software makers.

Not in this universe.

Here, the Electron was hit with a succession of body blows. First brought to a standstill by production problems with the crucial logic chip, then whacked by a weakening demand for home computers, before being brought to its knees thanks to Acorn suffering a financial crisis that threatened its very existence.

And it all seemed so promising. 'Speaking privately to Popular Computing Weekly,' the magazine whispered in its 6 May 1982 edition [1], 'Acorn director Hermann Hauser said that a new computer would be launched in the third quarter of this year – sometime between July and September. The new Acorn computer will probably be called the Electron and cost between £120 and £150... in effect the machine will be a miniaturised BBC Microcomputer with higher resolution graphics than those offered by the Spectrum.'

'We needed a cheap machine to compete with Sinclair and to do it better,' remembers Acorn co-founder Christopher Curry. Without BBC branding, there would be no licensing cost to pay. It could also slash costs by simplifying the electronics inside. Yet this new machine would be compatible with the BBC Micro, making it ideal for both the home and education markets as it would be so easy to run existing BBC titles.

Had the Electron appeared in late 1982, as Hauser predicted, then it would surely have been a huge success. People were steering clear of the Model A version of the BBC Micro and its limiting 16kB of RAM, but £399 for the 32kB Model B was too much for a Britain still emerging from a 1981 recession. When the average weekly wage before taxes was £99 for women and £154 for men, that's far more than most households could afford.

While the Acorn engineering team weren't fans of producing a computer that was effectively a step back from the Micro, they dutifully followed orders. Almost immediately, however, the Electron's development hit problems. One of Acorn's big cost-cutting measures was a huge reduction of the overall number of chips on the board: from 102 to around a dozen. Acorn could achieve this through improvements in the manufacturing of an Uncommitted Logic Array (ULA), to use the term favoured by Ferranti, Britain's chief maker of such chips.

Steve Furber, one of Acorn's key engineers in the design of the BBC Micro, was the logical man for the job. With a little assistance from a Cambridge University student drafted into help over the summer of 1982, Furber carefully designed the ULA's layout to avoid the overheating problems that had delayed the launch of the BBC Micro in late 1981. 'I was extremely careful in the design of the high-speed parts, to make sure that everything was entirely within spec, and we still had problems,' says Furber. 'We had a disagreement with Ferranti as to what the source of these problems was.'

In short, Ferranti blamed Furber's design; Furber blamed Ferranti's manufacturing process. 'In the end, I persuaded them to increase the logic swing on the chip by 50%, which was fairly easy by removing a few components at the edge, and lo and behold, when they did that, all the problems went away. And I felt thoroughly vindicated.'

While it's easy to explain these problems away in a paragraph or two, in reality it took months to solve the production challenges. Hauser's 'private' suggestion to Popular Computing Weekly that the Electron would go on sale in the late summer/early autumn of 1982 was never realistic, but Acorn did go on record as saying the computer would launch in March 1983. It finally made its public debut in August that year.

Acorn spared no expense at a glitzy launch in the ballroom of London's Park Lane Hotel. As John Caswell, Acorn's Marketing Manager, explains, he was keen to emphasise to the hundreds of press in attendance that this was a home computer. 'We had Wendy Craig, who was a famous actress in those days, and we built a home, literally a house, in the basement, in the big room in the Park Lane Hotel in London.'

The newly launched Electron User magazine was suitably glowing in its praise for the event, although its mansplaining approach looks dated now. After TV presenter Cliff Michelmore had welcomed attendees, Craig appeared in what Electron User described as 'a poor little housewife, baffled by all this, sceptical of the use of microcomputers in the home and not wanting to be blinded by science' [2]. Thankfully for the little lady, Michelmore was on hand to explain.

Ironically, when viewed through a 21st century prism, Christopher Curry would go on to advocate the Electron as a weapon in women's battle for equality at the launch. 'Girls are 13 times less likely than boys to use a micro at home, and only 4 percent of micro users are mothers,' he told the audience. 'We hope that the combination of the Electron's educational pedigree and its potential application in so many areas relevant to women will help to reverse this trend.' Almost as ironically, comedian Stanley Baxter would also record an advert where he dressed up in various female guises to explain the Electron's appeal [3].

At this point, everyone involved was confident that the Electron would be widely available by Christmas 1983. Acorn shipped review systems to the specialist computer magazines, most of which responded in glowing terms. 'The Electron is an excellent micro for the money,' wrote Neville Maude in the October 1983 issue of Practical Computing [4]. He went on to recommend it as a 'first computer on which to learn' and praised its keyboard, graphics, BASIC (now at BBC BASIC II) and 'strong connections in the education field'.

Steve Mann was similarly positive in the October 1983 issue of Personal Computer World, describing the Electron as 'one of the most impressive machines I have seen. I'll stick my neck out a bit here and forecast this one will be the machine to challenge the Spectrum.' [5] He went on to say that 'the Electron positively oozes quality', heaping particular praise on the BASIC, the graphics, the high maximum resolution, and the keyboard.

Not that everyone was convinced by the Electron's charms. In December 1983, the Guardian's Jack Schofield was actively discouraging people from buying the Acorn for Christmas 'even if you can find one' [6]. For him, there were just too many limitations: 'no joystick ports, no cartridge slot, no printer port, and no way of driving discs. Making up for those deficiencies will, in the long run, add considerably to the real cost of this machine.'

Jack was also wary of the graphics-based sacrifices Acorn had made to keep the cost down. Top of the list was teletext mode, or Mode 7. This enabled BBC Micros to show 40×25 characters in seven colours (eight if you include black), as used by BBC Ceefax. But it was also used by the BBC's own software packages for the Micro, and had the advantage of consuming a mere 1kB of video memory.

For memory was another one of the Electron's problems. If you wanted to use the Electron in its highest resolution of 640×256 with two colours, that would take 20kB of the available 32kB of RAM. With the operating system also needing access to memory, this left programmers with 5kB. Even in its lowest resolution mode, with 40 rows by 25 columns of white-on-black text, they only had 15kB to play with. Programmers used clever tricks to get around these limitations, but it was an undeniable obstacle.

Schofield's words of warning had little effect on demand, with WHSmith ordering 100,000 units. In its November 1983 issue, Electron User stated that the 'demand for the Electron is so intense that the production line in Malaysia, where they are working overtime to meet Acorn's order for 100,000 machines, cannot cope'. [7] The magazine reported that the first production models shipped in September, and were 'immediately snapped up by dealers and software developers'. But there was a word of warning for its readership still waiting for deliveries: 'it is unlikely sufficient machines will be available to meet all pre-Christmas orders'.

This turned out to be a huge understatement, with Curry − interviewed for this book in late 2019 − putting the problem squarely at Ferranti's door. 'Ferranti had problems. The big launch day came and we hadn't got a product and we lost a whole year when the Electron would, at that time, have completely wiped out all the competition. And in that year, the world moved on. By the time that Ferranti chip was working, the year was over and we had to start looking at that awful product from Alan Sugar [the CPC 464].'

Curry had good reason to detest the Amstrad rival. Its £229 incarnation not only came with 64kB of RAM, it also included a dedicated monitor and a built-in cassette drive. As a thing of beauty, it fell some distance behind the sleek Electron; as a thing of value, there was only one winner. And, far from dominating the sub-£200 market, by the time the Electron finally became widely available in mid-1984 it had tough rivals in the form of a cut-price Commodore 64 and, come October, the 48kB ZX Spectrum Plus − complete with injection-moulded keyboard.

This left the Electron in an awkward place. It definitely wasn't the best choice for gaming; unlike the Commodore 64, there were no hardware sprites to help programmers make games more dynamic with limited memory. As Jack Schofield pointed out, there was also no joystick port. And when it came to the portfolio of games available − unless you wanted to play *Elite*, it was Game Over when compared to the Commodore and Spectrum.

The Electron was also a non-starter if you wanted a word processor. Yes, you could buy a hardware add-on that would allow you to hook up a printer, but in 1984 that would cost between £40 and £50 for a third-party device – before you even bought the printer. And this was one of the common criticisms laid at the Acorn Electron: once you started expanding it with the features you wanted, you were halfway to the cost of the more advanced BBC Micro Model B.

Come Christmas 1984, Acorn had the reverse problem of the previous year: hundreds of thousands of Electrons sitting unwanted in warehouses. This oversupply was against the background of the past two Christmases where home computers had been the must-have purchase, and it's only in retrospect that we wonder why computer manufacturers didn't hear the music a little earlier. With the wisdom of that hindsight, we can now point to the fact that CD players had been a niche buy after their launch in 1983, but dropping prices and growing music catalogues made them an obvious choice for tech-lovers in 1984.

Meanwhile, Sinclair moved into attack mode. One of its December 1984 full-page adverts [8] hit the Electron with a double whammy: first it would give away six of its most popular software titles 'free' with the Spectrum 48K, including *Horace Goes Skiing* and *Scrabble*; second, it focused on usable memory. The Spectrum 48K offered 41.5kB for £130, the Commodore 64 provided 37.9kB for £199. The Electron, with its usable memory of 20.8kB, was starting to look unappealing at £199. While no computer sold in outstanding numbers over Christmas 1984, the Electron barely made a dent.

In January 1985, Acorn took the inevitable decision to cut its prices. It would go toe to toe with the now-reduced Spectrum Plus at £129, and hope that its growing selection of games and educational titles would attract home buyers who weren't tempted by games alone.

At these prices, Acorn was selling at a loss but at least turning some of its assets into cash. Unfortunately for a company that had haemorrhaged money in a valiant but doomed attempt to introduce the BBC Micro to the USA, at the start of 1985 it was in a horrible situation. 'It became clear in mid-January that we could not pay our bills,' said Dr Alex Reid, Acorn's new acting chairman and chief executive [9]. 'The shortfall in projected sales left us with large stocks in which our cash is tied up. The banks will lend money against debtors, but not against stocks.'

Acorn needed a knight in shining armour, and it came in the form of Olivetti's CEO. In the space of seven years, Carlo de Benedetti had transformed a company best known for typewriters into one of Europe's computing giants. Its stated aim was to be number one, ousting IBM, and buying a 49% stake in Acorn gave it a useful foothold into the UK. And all for £10 million.

It seemed like a good deal for Acorn too. 'Olivetti sold typewriters all over the world,' says Curry. 'Through the same distribution they were going to deal with this terrible unsold stock [of Electrons]. It looked like a very good arrangement. A year later, they hadn't sold anything.'

In a scene that Curry compares to a Martin Scorsese movie, this led to a showdown moment. 'So a year passed and Hermann and I went over to see Benedetti and said, "Well, what's going on? Why haven't you sold any of our computers?" He said, "I don't know, haven't we?" "You haven't sold one," we told him.'

Benedetti immediately summoned Olivetti's sales and marketing staff. 'We sat on this top table and about 200 people filled up this room,' says Curry. 'And he said, "Well, look, these gentlemen here are our partners and we own a big part of their company. Why haven't we sold any products?" And there's complete silence. And everybody looked around and nobody said anything. "So I ask again, why haven't we sold anything?"'

Still silence, so Benedetti bangs his fist on the table in true Italian style. 'He says, "Come on, I want to know. You!"' recalls Curry. 'And a reluctant senior sales and marketing person said, "Well, signor Benedetti, two years ago we sold a lot of stock to AT&T and we had an agreement. Part of our agreement is that we went from our typewriter business into the computer business, we abandon our own computer, and we only sold IBM-compatible PCs. If we sold anything else we'd be in breach of our agreements with AT&T."'

If Hauser and he were 'proper, sharp, hard, legalistic businessman' then they could have sued Olivetti for breach of contract, Curry now believes, but the end result was that the Italian company exercised its option to buy further shares and took outright control of Acorn. Within a year, they were slashing costs and shutting down the company's various projects.

'The only thing we didn't shut down was the ARM development,' says Curry. Which, as we shall see in the story of the Acorn Archimedes, turned out to be good news for Acorn, Olivetti, and anyone who's used a smartphone or Raspberry Pi in the past decade.

The Electron legacy

While history correctly judges the Electron as a failure – how can you argue when it lost Acorn several million pounds and was one of the key contributing factors to its eventual demise as a company? – it remains an important part of a lot of British people's lives.

Gavin Smith describes it as his 'first ever machine. It had a real keyboard, unlike the Spectrum. Taught me to type. Taught me to code. Taught me to explore computing.' Gavin now runs a software company that writes apps for iOS and desktop computers.

Another Electron fan prefers to stay anonymous, perhaps in case his or her Mum is reading. '[I remember] playing *Elite* whilst my parents thought I was learning how to write code. It turned out that "10 PRINT "Hello Mum" 20 GOTO 10" was enough to convince them I was a developing computer genius.' Our games-playing hero has gone on to work in academic libraries throughout his/her career, 'introducing IT into how we deliver information and services'.

'Hooked me for life on computers, as it was where I first learned and explored programming,' said another anonymous responder, although they have less reason to be embarrassed: 'I'm a software developer with a PhD in computer science.'

While the Electron doesn't have the same legacy as the BBC Micro, there is an argument that Acorn's loss was a gain for the country. It forced the company to release a computer that was a brilliant platform for learning how to program on, and who knows how many children were persuaded to start programming rather than fritter time away on a game, simply because there wasn't a compelling one for them to play?

What came next

While the Electron never had subsequent models – production was shut down in 1985 – Acorn did produce two interesting add-ons that took advantage of its expansion port.

Acorn Plus 1

Release September 1984   **Price** £60

This expansion module provided a parallel port for Centronics printers, plus two cartridge slots for adding compatible games, application or language ROMs, or additional expansion devices. It also gave users four 8-bit analogue to digital input channels, which meant you could add two joysticks or four games paddles.

Acorn Plus 3

Release March 1985   **Price** £229

We haven't forgotten the Acorn Plus 2: it was meant to add Econet but was never released. But if you wanted to add a 3.5-inch floppy disk drive to your Electron, you were in luck thanks to the Acorn Plus 3 – because that's precisely what it does. As well as being expensive, it was huge: a giant L that hooked all the way along the Electron's back and then extended it by a few centimetres to the side.

Sources

Interviews with John Caswell, Christopher Curry, Steve Furber, Hermann Hauser, and Jack Schofield.

1. Staff reporter, *Sinclair Spectrum stuns BBC*, Popular Computing Weekly, 6 May 1982, page 5
   archive.org/details/popular-computing-weekly-1982-05-06/page/n3/mode/2up
2. Staff reporter, *Well and truly launched*, Electron User, October 1983, page 3
   archive.org/details/Electron_User_Vol._1_No._1_1983-10_Database_Publications_GB/page/n1
3. Stanley Baxter, Acorn Electron computer advert, YouTube
   www.youtube.com/watch?v=kHmnilpSD64
4. Neville Maude, Acorn Electron review, Practical Computing, October 1983, page 68
   archive.org/details/PracticalComputing1983October/page/n65/mode/2up
5. Steve Mann, Acorn Electron review, Personal Computer World, October 1983, page 167
6. Jack Schofield, *Dreaming of a byte Christmas*, The Guardian, Thursday 8 December 1983, page 18
7. Staff reporter, *Production rate is doubled*, Electron User, November 1983, page 3
   archive.org/details/Electron_User_Vol._1_No._2_1983-11_Database_Publications_GB/page/n1/mode/2up
8. Sinclair ZX Spectrum advertisement, The Guardian, 22 December 1984, page 5
9. Stella Shamoon, *Acorn – an Italian connection*, The Observer, 24 February 1985, page 33
10. The Guardian (London edition), Tuesday 10 September 1985, page 23

Apple Macintosh

The computer that
started a revolution

Here's the popular history on how the Apple Macintosh was born. Steve Jobs visits Xerox PARC, the beating heart of computer innovation during the 1970s, and happens to see a graphical user interface and mouse for the first time. He cries 'Eureka!', gathers together the brightest Apple brains, and they beaver away until the historic, revolutionary Macintosh bursts into glorious being. In particularly glossy versions of history, the Mac is an instant hit, fighting The Good Fight against evil IBM and Microsoft.

There are certainly elements of truth in that. Yes, Steve Jobs visited Xerox PARC and saw a graphical user interface and the mouse. But by this point, November 1979, the Apple Macintosh project was already well underway, led by Jef Raskin – who, as you will soon find out, had a somewhat turbulent relationship with Jobs.

The true birth of the Macintosh project took place during a meeting between Raskin and Mike Markkula in March 1979, when Apple's chairman asked him to design Annie. 'It was supposed to be a $400 game machine,' said Raskin [1]. 'But I counter proposed, and said, "Well, I've been thinking about something I call Macintosh." It would give all the power of the computer, but with greater ease of use.'

Markkula asked him to develop those ideas, which Raskin did – in typically offbeat style – in a May 1979 document called 'Design Considerations for an Anthropophilic Computer' [2]. His vision, even then, holds echoes of what finally appeared almost five years later:

This is an outline for a computer designed for the Person In The Street (or, to abbreviate: the PITS); one that will be truly pleasant to use, that will require the user to do nothing that will threaten his or her perverse delight in being able to say: "I don't know the first thing about computers," and one which will be profitable to sell, service, and provide software for.

Where Raskin's vision begins to diverge from the final Macintosh comes shortly after: 'The computer must be in one lump. This means, given present technology, a 4 or 5 inch CRT (unless a better display comes along in the next year), a keyboard, and disk integrated into one package. It must be portable, under 20 lbs, and have a handle.' He goes on to speculate that it could even include a battery that would keep it running for at least two hours.

What isn't laid out at this early stage is a graphical user interface or a mouse. Indeed, key Macintosh software engineer Andy Hertzfeld states that Raskin was 'dead set' against the mouse, 'preferring dedicated meta-keys to do the pointing' [3]. Raskin always denied that he was anti-mouse, but appears to have preferred using 'meta-keys' and a graphics tablet for his own use.

Just like the Xerox PARC-inspired Sinclair QL, the idea of a bitmapped display emerged early on in the Macintosh's progress, and Raskin was certainly a keen advocate of computer graphics: 'On the Macintosh, the dreams I had about what I wanted to do on it all involved graphics,' he later said [4]. 'I wanted to be able to compose music on it, I wanted it to be able to handle musical notation, I wanted it to be able to handle pictures and photographs.'

Raskin was also determined to deliver the Macintosh on a tight budget, with his target price being $500. Even after this rose to $1,000, this put him in direct conflict with Steve Jobs, with Raskin delivering a biting response to the Apple co-founder's observation, 'Don't worry about price, just specify the computer's abilities.' Raskin sarcastically describes a printer that 'takes ordinary paper and produces text at one page per second' and declares this miracle computer 'can also synthesize music, even simulate Caruso singing with the Mormon tabernacle choir' [5]. Jobs' response is unknown...

Despite this, Raskin was keen to get Jobs on his side. Indeed, one of the reasons why Jobs visited Xerox PARC in December 1979 was because Raskin (using software engineer and shared friend Bill Atkinson, who was then working on the Lisa project with Jobs, as a conduit) encouraged him to do so.

Raskin's motivation was simple. Many of the innovations created by Xerox PARC were instrumental in his vision of the Macintosh, and he wanted Jobs to see them first-hand. In particular, Raskin believed in the importance of an intuitive user interface and – as a one-time visiting professor to the PARC during his previous career at the University of California – he was already familiar with the Xerox Alto. Created in 1973, this would be immediately familiar to a modern-day computer user thanks to its graphical user interface, mouse, and (though introduced later) desktop metaphor. For Xerox, the Alto was a research project, with around two thousand built but none sold.

Unbeknownst to Raskin, Jobs already had good reason to visit the PARC. Xerox had just invested $1 million in Apple shares, and part of the agreement was that Jobs could receive a tour of the PARC's restricted areas where its most cutting-edge ideas were being developed. It was here – on a second visit, and after the intervention of Xerox's venture capital arm – that Jobs saw a graphical user interface in action for the first time. To say he was impressed would be a mild understatement. 'It was like a veil being lifted from my eyes,' said Jobs [6]. 'I could see what the future of computing was destined to be.'

The visit gave Jobs renewed energy for the Lisa project. While Lisa is now of academic interest, in the early 1980s it was Apple's great hope to replace the Apple II and become America's de facto business computer. Its main innovation was a 16-bit processor, the Motorola 68000, rather than the 8-bit chip inside the Apple II, and with it the ability to address far more memory. Until Jobs's visit to Xerox PARC, however, he couldn't see how to make the Lisa truly innovative and revolutionary. Now, his path forward was clear.

Jobs used his legendary enthusiasm to motivate the Lisa development team, explaining that this easy-to-use computer would include both a graphical user interface (GUI, pronounced 'gooey') and a mouse. Despite this promising start, the project soon hit major bumps in the road. Jobs's instincts led him to prioritise the end-user experience, which was less important (or so everyone appeared to think at the time) for a business-focused computer. It didn't help that Jobs expected development to be far faster than was feasible: on seeing the Xerox Alto's interface, Atkinson had told him that development of Apple's own GUI would take six months. This proved wildly optimistic.

When his team missed the unrealistic deadlines he had set, Jobs hurled insults at them. He undermined the managers who he himself had hired to deliver the project. In September 1980, with Jobs's behaviour now actively disrupting progress, those managers decided enough was enough. They complained to Apple's president, Mike Scott, who decided to remove Jobs from the Lisa project.

Throughout this time, Raskin was working on the Macintosh project with the help of three others. A young engineer called Burrell Smith designed the first prototype board: this featured a modest Motorola 6809E processor and 64kB of RAM, and could power a 256×256-pixel mono bitmapped display. (For those who like ridiculous trivia, the first on-screen image it produced was of Scrooge McDuck, courtesy of Hertzfeld, in February 1980. [6])

With few resources, by autumn 1980 the Macintosh's momentum was stalling rather than building. The Apple board had already cancelled it twice, only to relent when Raskin begged for more time. Bud Tribble, who would later become Vice President of Software Technology at Apple, joined the project in September 1980 and persuaded Burrell Smith to create a new design around the Motorola 68000.

It's at this point that Steve Jobs, having nursed his Lisa wounds, returned to Apple and turned his attention to the Macintosh. In theory this was good news for Raskin, because with Jobs came a massive uptick in both attention from the board and investment. But, as Raskin's earlier barbed response to Jobs indicates, the divides in approach and philosophy between the two men were too great.

One obvious example was cost. Raskin's vision was for an affordable computer, which is why he backed the Motorola 6809E processor with 64kB of RAM. Jobs cared less about the final price of the machine and wanted the Macintosh to be based on the more expensive Motorola 68000, as used in the Lisa. This also meant the RAM had to double to 128kB. By January 1981, after enduring weeks of Jobs undermining him and his ideas, Raskin wrote a memo to Mike Scott entitled 'Working for/with Steve Jobs' [8]:

> He is a dreadful manager. … I have always liked Steve, but I have found it impossible to work for him. … Jobs regularly misses appointments. This is so well-known as to be almost a running joke. … He acts without thinking and with bad judgment. … He does not give credit where due. … Very often, when told of a new idea, he will immediately attack it and say that it is worthless or even stupid, and tell you that it was a waste of time to work on it. This alone is bad management, but if the idea is a good one he will soon be telling people about it as though it was his own.

To Scott's credit, he immediately acted on the memo and called both Jobs and Raskin to his office that same afternoon. Both sides pleaded their case, with Jobs breaking into tears. This time, the Apple president sided with the company's visionary co-founder and instructed Raskin to take a six-month leave of absence.

'There is no question that it was the right decision,' says Hertzfeld. 'Mike Scott thought Steve was more in his element leading the tiny Macintosh team than the large

Lisa team, and that he would have a far better chance than Jef at doing something significant. The Mac team was only five people at that point, and wasn't considered important to Apple, so there wasn't much to lose and a lot to gain if Steve pulled it off.'

Ultimately, the team also backed Scott's decision. 'There were mixed feelings because everyone liked and respected Jef, but realised that Steve was much better for the project,' Hertzfeld says. 'Everyone thought that the 68000 and mouse that Steve was championing was a better direction than Jef's ideas.' Plus, Hertzfeld describes the 'increased excitement to be working closely with Steve'.

Whatever Scott's motives, it turned out to be the best decision of his four-year tenure at Apple. This time Jobs was in charge of the right project, with his laser-like focus on creating the best possible product for consumers, rather than businesses, tying in with the Macintosh's goals much more than the Lisa.

He set to work immediately. He moved the Macintosh team from a small office to the two-storey 'Texaco Towers', and began to build up a fresh mix of newcomers and Apple's best talent. Among them was software engineer Andy Hertzfeld, who Jobs poached from the Apple II development team. When Jobs appeared at his desk to tell him the news, Hertzfeld said he needed a couple of days so he could hand over his DOS program for someone else to take over. 'Who cares about the Apple II?,' said Jobs [9]. 'The Macintosh is the future of Apple, and you're going to start on it now!' Jobs then pulled the power cord, causing all Hertzfeld's code to vanish.

We asked Hertzfeld how he felt at that point. 'I was a little bit pissed off that he pulled the plug, losing my recent work, but I was mostly astonished that he did that, and I didn't have much time to react – I had to follow him once he grabbed my Apple II,' he says. Jobs then drove Hertzfeld and his computer to Texaco Towers, spending the journey 'talking about how great the Macintosh was going to be, and how great the team was'.

In April 1980, while still working on the Lisa project, Jobs had commissioned Dean Hovey from nearby design firm Hovey-Kelly to work on the mouse. While its design may seem obvious now, the firm had to overcome many obstacles to meet the challenge of converting Xerox's expensive and unreliable three-button mouse into a commercial and robust version that would cost less than $35 per unit.

The decision to make it one button rather than two or three came easily, with Hovey citing simplicity as the main reason, but the collective wisdom of Apple and

Hovey-Kelly needed to solve many practical problems. How do you avoid the ball slipping? By creating a cage for it rather than relying on a bearing support. How do you make it easy to clean? By using a locking ring. Hovey-Kelly went through 20 different prototypes before they landed on a design that everyone was happy with.

The Macintosh's user interface went through similar iterations as the growing team of Macintosh engineers worked – often in tandem with the Lisa team – to improve on the Xerox Alto's GUI. It's key to note that the Alto inspired the Apple GUI, but it was far from a copy. 'We never had access to an Alto or Star or any of their software,' says Hertzfeld. It's impossible to underestimate the work Apple's team of software engineers did to take Xerox's GUI concept, from scratch, and turn it into a working and attractive environment.

Nor should we forget the limited hardware resources available. 'By far, the biggest challenge was getting everything to run well in only 128kB of RAM, as well as fitting all the code into 64kB of ROM,' says Hertzfeld. As the author of *Revolution in the Valley*, published by O'Reilly in 2004, he is also keen to point out that he probably gets more credit than is due for the creation of the Mac. So who else should be in the Macintosh hall of fame? 'Bill Atkinson was by far the most important developer, even though he wasn't officially part of the Mac team until 1983,' says Hertzfeld. 'He gets a lot of credit but deserves even more. Larry Kenyon also was a mostly unsung hero, making a tonne of crucial contributions, and Bruce Horn and Steve Capps deserve more credit than they usually get.'

Naturally, the other person who gets credit is Steve Jobs. While not a programmer, his attention to detail and relentless driving of the Macintosh team to produce the best possible product was crucial. Take rounded corners. In May 1981, Bill Atkinson had added routines to QuickDraw – the Macintosh's underlying graphics engine – so that it could quickly draw circles and ovals. He was excited, but Jobs wanted more. 'Well, circles and ovals are good, but how about drawing rectangles with rounded corners?' [10]

When Atkinson didn't respond, Jobs started pointing around the office at all the rectangles with rounded corners in real life. The desks, tables, whiteboards. He even dragged Atkinson for a walk around the block to convince him. 'OK, I give up,' said Atkinson. 'I'll see if it's as hard as I thought.' The following afternoon, he came back into the office with the completed code.

Jobs was equally demanding when it came to the hardware design. He even rejected the initial motherboard design for being too ugly and too congested, only relenting when his suggestions for a prettier design proved unworkable. Similarly, he wanted the Mac's rubber feet to have an Apple logo stamped into them, but this proved expensive and impractical.

There would be no such compromises with the case. This was the item with the longest lead time so needed to be completed first, with all the work taking place between March 1981 and February 1982. Jerry Manock and Terry Oyama worked together to create the prototypes, with Jobs having decided early on that the monitor should sit above the disk drive to minimise the footprint.

On seeing the first prototype, Jobs was typically cutting in his criticism. 'It's way too boxy, it's got to be more curvaceous,' he said [11]. 'The radius of the first chamfer needs to be bigger, and I don't like the size of the bezel.' Then came the bone. 'But it's a start.' Over the next few months, Manock and Oyama would come in with a revised version every few weeks, until Jobs finally signed off on the design and it was sent to Apple's industrial design team to turn into something that could be actually manufactured.

While development of the case appears relatively smooth, the development of the software continued to be a slow grind. To keep the momentum going, and the team spirit strong, Jobs organised six-monthly off-site retreats, starting in January 1982. 'A retreat usually lasted two full days, including an overnight stay,' wrote Hertzfeld [12]. 'We'd travel by bus to a naturally beautiful resort an hour or two from Apple's offices in Cupertino, like Pajaro Dunes near Monterey Bay. Every employee on the team was invited, as well as folks from other parts of the company who were contributing to the project.'

You can see the shifting priorities of Jobs through the aphorisms he used to summarise the agenda. In September 1982, they were: 'It's Not Done Until It Ships', 'Don't Compromise!', and 'The Journey Is The Reward'. In January 1983, just after Lisa launched, they were 'It's Better To Be A Pirate Than Join The Navy', 'Mac in a Book by 1986', and 'Real Artists Ship'.

And 1983 would prove to be the decisive year for the Macintosh. As the release date drew nearer, the team outgrew their building for the third time – they had moved from Texaco Towers to Bandley 4 on the main Apple campus in spring 1982 – and shifted to the larger Bandley 3 building. Inspired by Jobs's reference to pirates in that year's retreat, programmer Steve Capps decided they should fly a black pirate flag, with a skull, crossbones, and an Apple logo for an eye patch above the building.

The flag proved a potent symbol of the renegade mentality Jobs had imbued into the Macintosh team. And they would need it. During the autumn of 1983, Jobs committed to a launch date of 24 January 1984 – Apple's annual shareholders meeting.

There was much to do. With no hard disk option, the Macintosh was totally dependent on the floppy drive. The Lisa used two 5.25-inch Apple FileWare drives, nicknamed 'Twiggy' after the famously slim British model, but they were too unreliable and slow. Jobs preferred Sony's 3.5-inch floppy, but he insisted that Apple design their own version in tandem with Apple's preferred manufacturer, Alps Electronics.

The Macintosh engineers reluctantly agreed, but secretly carried on talks with Sony as they believed there was no chance that Alps could deliver a finished product in time. Their secrecy was such, and their fear of Jobs's reprisals so great, that when Sony sent a young engineer called (aptly) Hide Kamoto to discuss specifications, and they heard Jobs approach their desk, Apple engineer George Crow told Kamoto, 'Dozo, quick, hide in this closet. Please! Now!' [13] All this effort paid off, however, with Jobs agreeing to use Sony's drive when Alps came back with an 18-month lead time for their version of the 3.5-inch drive.

'As pressure mounted to finish the software in time to meet our January 1984 deadline, we began to work longer and longer hours,' wrote Hertzfeld [14]. 'By the fall of 1983, it wasn't unusual to find most of the software team in their cubicles on any given evening, weekday or not, still tapping away at their keyboards at 11pm or even later.'

New Year quickly rolled around and the deadline loomed ever larger. 'By the first week of January, the software team was working around the clock, testing and fixing problems that were found,' remembered Hertzfeld [15]. 'Every employee in the building was drafted as a tester.' The finance department even decided to commemorate the effort by creating a grey hooded sweatshirt with '90HRS/WK AND LOVING IT' emblazoned on the back.

Despite this Herculean effort, the job was still far from done on Friday 6 January, with just over a week to go before the software needed to be shipped to the factory. The team begged Jobs for an extra week to finish it, but the answer was typically robust: 'No way, there's no way we're slipping!', he said. 'You guys have been working on this stuff for months now, another couple of weeks isn't going to make that much of a difference. You may as well get it over with. Just make it as good as you can. You better get back to work!'

With an absolute deadline of 6am on Monday 16 January, the team worked all hours for the next week, including a sleepless final weekend fuelled by a combination of espresso beans and, in Hertzfeld's words, 'medicinal quantities of caffeinated beverages'. By 5.30am, they finally created a stable version they were happy with, and drove it to the factory.

Jobs arrived in the office at 8.30am to find Hertzfeld and Donn Denman, creator of MacBASIC, sitting zombie-like in the lobby. Hertzfeld brought Jobs up to speed before heading home to collapse. By 5pm he and the rest of the team were back in the office, anxious to make sure that the release was still on schedule. But there was no time to rest, as Jobs then issued the command that they needed to program a demo for the first showing of the Macintosh in public.

With the panache that would soon become something of a trademark for both Jobs and Apple, they decided to put on a show that took full advantage of the Macintosh's capabilities. First, Jobs lifted the Macintosh from a bag before plugging it in, switching it on, and inserting a floppy drive.

Cue the *Chariots of Fire* music and huge letters scrolling from right to left, spelling out 'MACINTOSH'. Then a skywriter writing 'Insanely great' in cursive script against a background of twinkling stars. Then screenshots of MacWrite, Program, and a selection of third-party applications that had been created for launch, including Microsoft Multiplan (a spreadsheet program) and Chart.

With this display of power complete, Steve Jobs announced: 'Now we've done a lot of talking about Macintosh recently. But today, for the first time ever, I'd like to let Macintosh speak for itself.' [16] The stage went dark and Jobs pressed the mouse button, at which point the Macintosh uttered these words in a robotic, synthesized voice:

Hello, I'm Macintosh. It sure is great to get out of that bag!
Unaccustomed as I am to public speaking, I'd like to share with you a maxim I thought of the first time I met an IBM mainframe: NEVER TRUST A COMPUTER YOU CAN'T LIFT!

Obviously, I can talk, but right now I'd like to sit back and listen. So, it is with considerable pride that I introduce a man who's been like a father to me... STEVE JOBS!

To say the shareholders gave the Macintosh an enthusiastic welcome would be underselling it. A beaming Jobs watched on as many stood up to applaud; one particularly enthusiastic lady even jumped up and down.

What those watching didn't know was that this big launch, and the advertising blitz that followed, came at a cost. Sculley insisted that, to pay for it, Apple would have to increase the price of the Mac from $1,995 to $2,495. This was some distance from the $1,000 target price of the Macintosh when Jobs took over the project; adjusted for inflation, it's around $6,500 in today's money.

This didn't stop Apple fans from buying the Macintosh, but by the end of 1984 sales had slumped to fewer than 10,000 per month. While Jobs blamed Sculley's decision to raise the price, the truth was that the first iteration of the Macintosh was underpowered. For example, to run the launch demonstration they had to install 512kB of RAM rather than the 128kB it came with.

And this is where Apple's sealed box approach also proved problematic. While it was technically possible to expand the memory to 512kB yourself (when an article published in the January 1985 edition of Dr Dobbs, entitled *Fatten Your Mac*, explained the process, the issue sold out), this was no trivial task. And with no hard disk inside the case, and just one built-in floppy drive, making a backup involved removing the diskette many times or buying an external floppy.

As the months rolled by, Apple even began to lose its marketing touch. It had changed the world of advertising in January 1984 when it teased the launch of the Macintosh with a one-minute ad during the Super Bowl. In the dystopian future it depicted, the mass populace (all skinheads) are in thrall to a Big Brother-like screen; an athletic young woman, dressed in bright running gear and wielding a sledgehammer, runs into the hall. She hurls the hammer into the screen, destroying it, at which point these words appear on-screen: 'On January 24th, Apple Computer will introduce Macintosh. And you'll see why 1984 won't be like 1984.'

Directed by Ridley Scott, and filmed in England using authentic skinheads who quite scared the on-set Apple executives, the 1984 ad went on to win many awards and is considered a watershed event in American advertising. But when Apple attempted to repeat the feat in 1985 with an ad to launch Macintosh Office, this time depicting blindfolded businesspeople walking off a cliff to their deaths as they do 'business as usual', it flopped. The very people that Apple were trying to attract were being told they were stupid.

In the first quarter of 1985, Apple posted a loss for the first time. To anyone apart from Jobs, the problems besetting the Macintosh were obvious: it needed more RAM; there wasn't enough third-party software available; and it desperately needed a hard disk. Jobs, though, was living in his own reality distortion field. He carried on using sales forecasts that were now quite obviously wrong, and no one was brave enough to challenge him for fear of one of his legendary outbursts.

When he heard that John Sculley – who Jobs had famously lured to Apple as president with the line 'Do you want to sell sugar water for the rest of your life, or do you want to come with me and change the world?' [17] – was planning to remove him from the Macintosh team and strip him of his management role, Jobs set up a rival plan. He would remove Sculley.

This led to an emergency board meeting on 24 May 1985. The attendees had to make one decision: either Sculley would go, or Jobs would be stripped of his responsibilities. While the decision was painful, it was also obvious: they opted for mature leadership over Apple's unpredictable but visionary founder.

Was it the right decision? History suggests yes. It allowed Sculley to put Apple back on secure foundations, releasing commercially successful versions of the Macintosh and removing some of the divisions within the company; Jobs had long denigrated the work of the Apple II team, making them feel like second-class citizens despite their key part in generating the firm's profits for almost a decade.

You can even argue the decision did Jobs good. When he opted to leave – he wasn't fired by the board, just stripped of responsibilities – and set up NeXT, he must have also felt a sense of release. Apple had grown from two geeks in a garage to a company with a $1.5 billion turnover in less than a decade. That's quite a responsibility.

As for the Apple Macintosh: it is, without doubt, one of the most important computers in history. 'We had high ambitions for the Mac, but we exceeded them, making the first computer that ordinary people could afford and enjoy using,' Hertzfeld tells us. 'It's almost certainly the most important computer of the 1980s and arguably the most important ever, at least before the iPhone. Most computers today, 36 years later, are based on the paradigms created and popularised by the Macintosh.'

Compared to everything that came before, it thought different.

If Jef Raskin had been left in charge...

You could argue that we saw two realisations of Raskin's vision of the Macintosh. The first came in the form of the Osborne 1, a 10.7 kg portable computer released in April 1981 that closely matched Raskin's early concepts: opening up its hinged lid revealed a built-in keyboard, 5-inch mono screen and two 5.25-inch floppy drives.

Where it differed was its specification and operating system, with Osborne Computer Corporation choosing a Zilog Z80 processor, 64kB of RAM, and Digital Research's popular CP/M operating system. With a hefty package of bundled software (WordStar, SuperCalc, and BASIC) and a retail price of $1,795, it was an instant hit, with 11,000 units shipped during the first eight months, with a further 50,000 on back order.

Sales dried up the following year, due in part to the company announcing the Osborne 1's successor – the Executive, with a more usable 7-inch CRT – long before it was ready to ship. New rivals soon appeared and the company ran out of money, filing for bankruptcy in September 1983.

In 1987, Jef Raskin finally brought out his own computer: the Canon Cat. While it looks toy-like now, Canon targeted secretarial workers with this all-in-one machine – not only was the 9-inch black-on-white screen built-in, along with a 3.5-inch floppy drive and even a modem, but so was the keyboard.

Notably, there was no mouse – ironic, bearing in mind the computer's name – with Raskin instead using two 'Leap' keys to help people jump to any word, sentence, or paragraph in the document they were working on. When users finished, they could save the document to floppy drive, send the text to another Cat computer, or print it out via the optional printer.

It was a powerful machine thanks to a Motorola 68000 processor and 256kB of RAM, but with a proprietary operating system and no graphics program – despite the Cat's inherent graphical abilities thanks to a bitmapped screen – it failed to sell.

Raskin later wrote: 'Canon did not want to reveal that it was actually a 68000-based bitmapped product with a nice set of graphics tools in ROM – which were tools they never used. This was partly because they decided to bundle it with a daisy-wheel printer (!) that could not do graphics, and partly because it was brought out by their electric typewriter division and not their computer division.' [18]

After six months, Canon had only shipped 20,000 units and the Cat, along with Raskin's dream of creating a computer for the 'average person in the street', was put to sleep.

Microsoft and the Macintosh

Apple and Microsoft always had an uneasy relationship, as echoed by that between Jobs and Bill Gates. Jobs knew that he needed Microsoft: it was the first third-party company that he recruited to create software for the Mac. But he was also wary of the firm as a competitor, and Gates as a business rival.

This is why, as part of the deal that gave Microsoft access to the Macintosh project's secrets so it could write software, he added a condition that Microsoft could not release an operating system that used a mouse for at least a year after the Macintosh started shipping. At the time of the agreement, signed in late 1981 when Apple expected the Mac to ship in autumn 1982, that date was defined as September 1983.

Microsoft duly started work on its Multiplan spreadsheet and Chart graphics applications (along with a word processor, but as that would compete with MacWrite it was kept a secret from Apple), with Microsoft's chief systems programmer Neil Konzen keeping in regular contact with Andy Hertzfeld from Apple.

When Konzen started asking questions about details that had nothing to do with the software Microsoft was writing, Hertzfeld got suspicious. He shared those suspicions with Jobs and others, but they weren't convinced. Then, in November 1983, Microsoft announced that it was producing a mouse-based GUI called Windows.

So did Hertzfeld feel betrayed by Neil Konzen? 'I didn't feel betrayed because it wasn't a surprise,' says Hertzfeld. 'If anything, I felt vindicated, because I was telling Steve and others I thought they were cloning the Mac for at least six months before the Windows announcement.

Jobs appears to have been more taken aback. 'Get Gates down here immediately!', Hertzfeld quotes him as shouting [19]. 'He needs to explain this, and it better be good!' The following day, Jobs confronted Gates in an Apple conference room. Gates arrived alone, while Jobs included several other Apple employees including Hertzfeld – who remembers the confrontation like this:

"You're ripping us off!", Steve shouted, raising his voice even higher. "I trusted you, and now you're stealing from us!"

But Bill Gates just stood there coolly, looking Steve directly in the eye, before starting to speak in his squeaky voice.

"Well, Steve, I think there's more than one way of looking at it. I think it's more like we both had this rich neighbour named Xerox and I broke into his house to steal the TV set and found out that you had already stolen it."

When Microsoft shipped Windows 1.0 in late 1985, it was still some way off the Macintosh's elegance. 'I was sort of surprised they did such a poor job of it,' says Hertzfeld. 'Windows 1 was ugly and barely worked and amazingly didn't even have overlapping windows. When I saw Windows 2, which slavishly copied the Mac including all of the desk accessories, I was surprised at their lack of imagination. It took them over six years to make a passable imitation with Windows 3.1 and even that wasn't close to the quality of the Mac.'

Apple still filed a copyright lawsuit, in 1988, but it ultimately lost its case. 'Apple cannot get patent-like protection for the idea of a graphical user interface, or the idea of a desktop metaphor which concededly came from Xerox,' stated the judgement in September 1994 [20]. 'It can, and did, put those ideas together creatively with animation, overlapping windows, and well-designed icons; but it licensed the visual displays which resulted.'

Sources

Email interview with Andy Hertzfeld.

1. Making the Macintosh: Technology and Culture in Silicon Valley, Interview with Jef Raskin, 3 June 2000
    web.stanford.edu/dept/SUL/sites/mac/primary/interviews/raskin/trans.html

2. Making the Macintosh: Technology and Culture in Silicon Valley, Design Considerations for an Anthropophilic Computer, May 1979
    web.stanford.edu/dept/SUL/sites/mac/primary/docs/bom/anthrophilic.html

3. Andy Hertzfeld, *The Father Of The Macintosh*, Folklore.org, Undated
    folklore.org/StoryView.py?story=The_Father_of_The_Macintosh.txt

4. Making the Macintosh: Technology and Culture in Silicon Valley, Interview with Jef Raskin, 3 June 2000
    web.stanford.edu/dept/SUL/sites/mac/primary/interviews/raskin/trans.html

5. Making the Macintosh: Technology and Culture in Silicon Valley, Reply to Jobs, and Personal Motivation, circa 1979,
    web.stanford.edu/dept/SUL/sites/mac/primary/docs/bom/motive.html

6. Walter Isaacson, *Steve Jobs: The Exclusive Biography*, Kindle edition, location 1938

7. Andy Hertzfeld, Scrooge McDuck, Folklore.org, January 2004
    folklore.org/StoryView.py?project=Macintosh&story=Scrooge_McDuck.txt

8. Walter Isaacson, *Steve Jobs: The Exclusive Biography*, Kindle edition, location 2173

9. As above, location 2196

10. Andy Hertzfeld, *Round Rects Are Everywhere!*, Folklore.org, January 2004
    folklore.org/StoryView.py?project=Macintosh&story=Round_Rects_Are_Everywhere.txt

11. Andy Hertzfeld, *More Like A Porsche*, Folklore.org, January 2004
    folklore.org/StoryView.py?project=Macintosh&story=More_Like_A_Porsche.txt

12. Andy Hertzfeld, *Credit Where Due*, Folklore.org, January 2004
    folklore.org/StoryView.py?project=Macintosh&story=Credit_Where_Due

13. Andy Hertzfeld, *Quick, Hide In This Closet!*, Folklore.org, January 2004
    folklore.org/StoryView.py?project=Macintosh&story=Hide_Under_This_Desk.txt

14. Andy Hertzfeld, *90 Hours A Week And Loving It!*, Folklore.org, January 2004
    folklore.org/StoryView.py?project=Macintosh&story=90_Hours_A_Week_And_Loving_It.txt

15. Andy Hertzfeld, *Real Artists Ship*, Folklore.org, January 2004
    folklore.org/StoryView.py?project=Macintosh&story=Real_Artists_Ship.txt

16. Steve Jobs introduces the Macintosh, Youtube.com,
    youtu.be/2B-XwPjn9YY, 22 January 1984

17. Walter Isaacson, *Steve Jobs: The Exclusive Biography*, Kindle edition, location 2847

18. Jef Raskin, The Creation of the Mac according to Raskin & Horn, Starway.org, undated
    starway.org/~arnaud/Raskin

19. Andy Hertzfeld, *A Rich Neighbour Named Xerox*, Folklore.org, January 2004
    folklore.org/StoryView.py?project=Macintosh&story=A_Rich_Neighbor_Named_Xerox.txt

20. Apple Computer v. Microsoft Corporation, Justia, 19 September 1994
    law.justia.com/cases/federal/appellate-courts/F3/35/1435/605245/

Amstrad CPC 464

Borrow boy to
computer king

Everyone in Britain knows the Alan Sugar story. He's the embodiment of a working-class kid made good, sticking it to the man with his straight-talking, no-nonsense approach. For once, this reputation matches the reality, because that's precisely what Lord Sugar – he was created a life peer in 2009 – did throughout his career.

But that simplifies things, because Alan Sugar's success was built on flexibility. In the late 1970s, he saw a way to supply hi-fi separates for a third of the price of high-street rivals. He seized on the CB radio opportunity when they exploded in popularity, and escaped the market the moment he realised that the bubble had burst.

Nor did he allow fate to take control. Guided by clear business principles, he was always on the lookout for ways to simplify the buying and setup experience for the general public.

By the end of 1982, with his company Amstrad riding high, Sugar was looking for new opportunities. And while he had initially dismissed the personal computer craze as a fad, when he saw the Sinclair ZX81 fly off the shelves, and the Commodore 64 follow suit, he knew the time was right for the nascent computing industry to get the Amstrad treatment.

Lord Sugar puts it succinctly in his autobiography, *What You See Is What You Get*: 'Ignoring all the high-tech bullshit spouted by the nerds who were seen to be the pioneers of the industry, we looked at how much this lump of plastic and silicon would cost to make.' [1]

He was none too impressed by the Sinclair ZX81, which 'looked like a pregnant calculator; it didn't look like good value for money. Also, at the time, people would buy a Sinclair computer and then have to buy a separate cassette player to connect to it. On top of that, they'd have to wire it all to the back of their television sets, which they'd use as monitors.'

All this broke Sugar's golden rules. Were Sinclair computers simple to use? No, not for the ordinary person on the street. Did they look both attractive and good value for money? Not in Alan's view: one of the hallmarks of the Amstrad 'tower system', an all-in-one hi-fi unit that earned the company millions in the early 1980s, was that it looked stylish but was also packed with buttons to give buyers the feeling that they were getting their money's worth. To put it in his own words, 'a mug's eyeful'.

Sugar was also determined that his computer wasn't going to sit on a dusty shelf. One problem with existing computers was that they monopolised the TV screen, so as soon as Dad came back from his day's work the kid would be kicked off. By integrating a TV, Amstrad wouldn't just solve that problem but make the computer so obvious in a room that it couldn't be ignored.

While Sugar was certain Amstrad could make a difference, many observers were less convinced. 'The industry was full of snobs who spoke in haughty, intellectual terms, trying to imply that the electronics involved in computers was something way above that used in the general consumer electronics industry,' wrote Sugar in his autobiography [2]. 'Fortunately, I recognised at a very early stage that this was a load of bollocks.'

Together with Bob Watkins, his trusted lieutenant and long-time partner for all things hardware design, they created a mock-up sample of hardware. There was just one problem: the internals needed to run it. In short, the Amstrad men had no idea how tough this was to create. In their innocence, they asked a couple of 'long-haired hippies' (Alan Sugar's words [3]) who had previously helped them with a technical project to produce the internal design and software.

Those hippies said yes; sadly, they were hugely underqualified to create the guts of a computer, never mind the software to run on it, and cracked under the pressure. This left Amstrad with a problem. It was committed to producing a mass-market computer by spring 1984, which meant delivering final hardware designs by the middle of December 1983. And the software to run on it not long after.

By now, it was late August. The company essentially had twelve weeks to create a computer from scratch, and as Amstrad had already committed to the physical design of the machine, the internals had to fit. Luckily, Bob Watkins knew just who to talk to: William Poel, who ran an electronic supplies shop and mail order company, and who Watkins had worked with successfully in the past when developing hi-fi equipment.

It's a meeting that has become legend among Amstrad folklore. Bob Watkins rolls up outside Ambit International's modest and unassuming shopfront in Brentwood, Essex, and not only was Poel available for a chat but so was a man who would become pivotal to the CPC's success: Roland Perry. In the middle of a stock-take, both men were bored and more than open to interruptions.

At this point, we should freeze time to take stock of Roland Perry himself. A Cambridge graduate, he'd just finished a major piece of work for Ambit: to create, from scratch, a computerised stock control system that would allow people to order a product (either in the shop or by post) and for the shop assistant to look it up on the system, see how many were available, log the sale, adjust the stock numbers, and create the sales invoice. It's essentially the same streamlined approach that has allowed Argos and Screwfix to become so popular, with the big difference being that Ambit rolled out its system in 1983.

No surprise, then, that Perry was growing bored of counting transistors as part of a stock-take on a slow August bank holiday weekend. And that's the point when Bob Watkins walked into the shop, carrying a cardboard box. Out of which he produced a keyboard with the letters 'AMSTRAD' writ bold in the corner. 'He just simply turned up and says, you know, what about this then?' remembers Perry. 'Lots and lots of strange things aligned. If they had told us, or we had realised at the time, how much work it was going to be, we would have just said "No thanks". It sounded challenging but deliverable. And so we said, "Oh, what the hell then, we're not doing anything else at the moment. So let's just dive in the deep end and do this." It turned out to be much more complicated than we thought.'

The complication, explains Perry, wasn't so much due to the design of the computer – something he had been doing for a decade – but all the 'frilly stuff'. Like having to write a user manual for it and persuading other people to write games for it. 'And I think part of the success of it was that we actually said, although we've only been commissioned to do the core design of this, actually we'll take on the frilly stuff as well. Without the frilly stuff it probably wouldn't have been such a success. They'd have only made a hundred thousand for a year and then, you know, gone off and done microwave ovens or something the next year instead.'

As Watkins drove away, Roland Perry had only vague notions of all that was to come. For the moment, he set himself to work. And within ten days, brilliantly, he had not only found a team of people to design the electronics and write the software, but created a 32-week plan that would take the CPC 464 from hardware shell to mass production.

A crucial part of the success of the CPC 464 was due to Perry's detailed plan and the fact he kept to it. But it was also due to the unwavering clarity of Alan Sugar's vision, crystallised by Bob Watkins. In fact, if you have any preconceptions about what Sugar is like as a businessman – the barrow boy image doesn't do him any favours – then Perry insists you should throw them away. 'He's just very straightforward', says Perry. 'He won't come back halfway through and say, "Oh,

I've changed my mind about this, or I'm not going to do it after all, or I want to renegotiate your fee, or anything at all". And that's just extraordinary. I've never worked with or for anybody in all the 35 years since, or the ten years before that, who's so straightforward.'

But Perry is quick to concede that the success of the CPC 464 was also due to an awful lot of luck. 'I've done dozens of projects that are superficially much the same [as the CPC 464], but they've all ultimately failed because something didn't click and then it didn't happen. What was good about the Amstrad project is that everything clicked.'

The first click: finding someone to take charge of the software. Perry approached Howard Fisher, an old school friend who was then working at Acorn. 'I went to him and said, "Oh, I've got this computer project which needs some software, in particular a BASIC interpreter. I know you can't do it, but is there anybody else you might know?"' As luck would have it, Fisher did: a chap called Richard Clayton from Locomotive Software.

Perry's next step was to meet Clayton, but don't imagine a glass-windowed office with a company logo visible for miles around. At that point, Locomotive's two employees – Clayton and Chris Hall – were working out of a spare room in Clayton's house. Perry removed the box to show Clayton the design, but by this time had painted over the AMSTRAD logo on the top-left of the keyboard because he was under strict orders from Bob Watkins that no one was to know the true identity of the company behind this computer (see 'The ARNOLD mystery').

To get around this, says Perry, he explained he was 'working for a large UK consumer electronics company. I told them, "I've got this home computer, it's going to be entry level, it's going to be better than a Spectrum. Hopefully, a little bit better than a Commodore 64."'

However, he hadn't reckoned with Richard Clayton's engineering instincts. 'He of course immediately opened up the box,' says Perry. 'At which point there's still an "Amstrad" written on the circuit board of the cassette recorder because Amstrad was very protective of its intellectual property; it wrote Amstrad on everything it did.' But here's another one of those lucky breaks. 'Luckily, he had never heard of Amstrad – or if he had, it didn't register. Or thought, "Oh, this must just be a cassette mechanism, from some Japanese company I've never heard of called Amstrad".'

Time for another cog to click into place. While Clayton could help with the software, he wasn't an electronics engineer, so couldn't design the circuitry. But it just so happened he was already working with someone who did: Mark-Eric Jones, who everyone called Mej. Mej even named the company he set up with a friend, Roger Hurrey, MEJ Electronics. MEJ and Locomotive had been working together on a project for Data Recall, a company owned by Mej's father, so knew each other well.

It wasn't long before Mej and Roger joined Perry, Clayton, and Hall in their front room. While there was a sense of excitement about this project, there were also problems. When Bob Watkins walked into the Ambit International shop with his fateful box, he also included a bill of materials: a set of components that Amstrad had already committed to buy. One of those components was a 6502 processor, because that's what the 'long-haired hippies' had said was needed. And to an extent, that made sense. It was the chip that powered the Commodore 64, after all.

Unfortunately, Clayton estimated, to create an operating system and BASIC from scratch for the 6502 would take him eight months. Perry clearly remembers the breakthrough moment when Clayton suggested switching to the Z80: 'He said, "well, we've got a BASIC we could use, but it's Z80 and there won't be time for us to convert it to 6502 so it's Z80 or nothing." That was the real killer.'

Interviewed for this book, Mej notes that the Z80 had other advantages too. 'Firstly, the Z80 managed the refresh on memory, whereas the 6502 didn't. And perhaps more importantly, the project that we had been doing for Data Recall was Z80-based and we all had a lot of experience of the Z80, both from a hardware and software point of view. I knew we could come up with a much simpler, cheaper design.'

Over the course of a pub meal later that night, they hatched a plan. If Amstrad was willing to switch from the 6502 to the Z80, then Mej could complete the hardware design within six weeks while Locomotive could deliver the operating system and BASIC by the end of January. This would give MEJ Electronics time to create prototypes, a sample gate array (more of this later), and test units.

All Perry had to do now was convince Amstrad that the Z80 was the right choice. Time to make a phone call to Bob Watkins. 'I'm sure he would have said to me, "Nah, I've already ordered one hundred thousand 6502s, you gotta use them. That's no good." And I'd say, "Come on Bob, you haven't really ordered them have you? You just want to keep me on the straight and narrow." But that sort of conversation didn't happen very often.'

With this one change to the original specification agreed, and Amstrad keen to keep things moving at speed, Bob Watkins arranged a meeting at Amstrad's Tottenham headquarters with Locomotive Software (Chris Hall rather than Richard Clayton, because, says Perry, he 'had a suit'), Mej, Roland Perry, William Poel, and one short, bearded, mystery gentleman.

Perry remembers the meeting with a laugh in his voice. 'Chris Hall recounts this tale that we went to the conference room, in Tottenham, and all the discussions were with Bob Watkins, because it was Bob Watkins's responsibility to rescue this project, to make sure that it continued. And there was this grumpy little bloke sat at the other end of the table who didn't say very much. Who turns out to be Alan Sugar, who it transpired was actually running the meeting, even though Bob Watkins was doing 99% of the talking.'

Mej was left equally in the dark, with no proper introduction to the 'bloke in jeans who apparently had just got off a plane from Asia'. This meant he didn't realise that Alan was the boss until partway through the meeting. The tell? 'Oh, the fact that everyone shut up when he started talking was a giveaway.'

But Mej remains full of respect for the business tycoon's way of working and the Amstrad culture as a whole. 'They were very quick at making decisions. They had this consumer product mindset, so they knew that you couldn't delay: it was better to make the wrong decision quickly, and then change it, than to analyse too much. And that was quite refreshing.'

For instance, Mej had realised that he could design a board using far fewer chips by using an ASIC (application-specific integrated circuit). At the time, these were called gate arrays or Ferranti ULAs, standing for Uncommitted Logic Arrays. Essentially, they're custom-made chips built for a specific purpose.

The only problem was that Amstrad's bill of materials didn't mention such a circuit. Fortunately, Mej had an argument that was right up Alan Sugar's street: more features, the same amount of money, and you won't be breaking your agreement with the suppliers… you'll just be ordering different components.

With the plan agreed, it was time to move into phase two of the project. While Locomotive was creating the software, Mej was designing the ASIC. Unfortunately, this wasn't a simple process. 'It was early days for the ASIC industry,' he remembers. 'Let's put it this way: we had some interesting challenges with the manufacturer.'

Here, the manufacturer was Ferranti, maker of electrical equipment since its foundation in 1885. It had since grown to be the pre-eminent name in UK electronics manufacturing, so was the obvious choice to create the ASIC specified by Mej. Unfortunately, this integrated circuit proved to be a huge challenge. 'It got to the point where I was going up to Gem Mill, which is where Ferranti had their operation for ASICs, so often that I was even recognised by a taxi driver at the airport.'

Mej points out that these were very different times. While today firms have computer-aided design and a bunch of tools to help deliver products quickly, in the early 1980s they were operating almost in a vacuum. 'At one point, even the salesman from Ferranti said, "Well, you could use the simulator but in practice it's going to be quicker if we just build some devices for you and you try them."'

After weeks of delays and excuses, Sugar was getting impatient. With the chances of the CPC 464 shipping on time starting to fall away, he demanded to know if anyone other than Ferranti could build these chips. His autobiography reveals he was none too impressed when told that actually some Japanese companies could. 'After letting off steam,' he wrote, euphemistically, 'I told them to ditch Ferranti and start again with a reliable Japanese maker.' [4]

Fortunately, the rest of the board design progressed relatively smoothly, with Mej relishing the challenge created by the fact that the CPC 464's design concept was already in place. 'These were interesting design constraints. Roger Hurrey came up with a very clever idea about how to implement more colours than you would normally get in a computer of that ilk. And then we built the gate array to do that.'

His involvement didn't stop once the design had been finalised, either, with Mej even employing the help of his then-girlfriend – now his wife – to create 40 prototypes in his front room 'It was a lot of fun, a lot of late nights, and a lot of frozen pizzas.'

But here was another problem. How could Perry and his team supply working prototypes of the CPC 464 to software houses if the gate array didn't yet exist? The answer: a GAS board. Perry explains: 'This was a gate array simulator, which was a massive daughterboard that allowed us to ship prototypes of the 464 before even the first sample gate arrays were made. The pizza-fed project at Mej's place was just as much designing and hand-building some of those, as the rather sparser main board itself.

'And of course,' he continues, 'by implementing the design in traditional logic chips like that, it proved the theory of the gate array design was correct. All that the chip manufacturers needed to do was implement it properly!'

Equipped with these prototypes, it was time for the third phase of the project as Perry set out on a charm offensive: attempting to persuade software developers to invest their time to produce software in time for launch. Games in particular.

This wasn't a simple sell. In late 1983, there were already numerous computer platforms that software developers could write for. To persuade them that this new kid on the block was worth spending time on, rather than creating something for the big hitters of the Spectrum 48K and Commodore 64, took a huge amount of time, spirit and, says Perry, preparation.

'When they first saw the prototypes, they were almost universally amazed at how professionally it was being presented to them. We wrote a manual in a ring binder, with a whole ream, 500 pages, of documentation about how the operating system and BASIC interpreter worked.' What's more, Perry presented each software house with its own pre-production prototype to develop on.

Then there was the practical side of things: to keep things as simple as possible for the developers. 'We'd gone to some trouble with the engineering of it,' says Perry, right down to the choice of individual components. 'We used the same chip to control the screen as other computers used, we used the same sound chip that other ones used, so that they didn't need to learn how to program a new chip.'

But one of the big advantages of the CPC 464 compared to the Commodore 64 and the Spectrum 48K is that it included 27 colours rather than 16 (kudos once more to Roger Hurrey here). 'And it was 27 real colours, not 16 colours which were the same eight in either dim and bright. That's a con,' says Perry. 'So they were impressed by that and thought there's something going on here that we can relate to.'

About a month after the initial rollout of prototypes, the software houses were invited to a hotel in Heathrow for a conference – a highly unusual step at the time. However, it allowed Perry and his team to 'give them a pep talk' and for Amstrad to go through its sales forecasts. This was also where he introduced the idea of Amsoft, a separate division within Amstrad that was dedicated to creating its own software for the CPC 464, along with fan magazines and a user club.

How key was Amsoft to the CPC 464's success? 'Absolutely essential,' insists Perry. 'Maybe the one thing we managed to persuade Amstrad that they hadn't spotted on day one was that they needed an Amsoft. They couldn't just send the computer off to Dixons and say to Dixons, maybe you can talk to this list of people for some aftermarket stuff. Dixons wanted to sell software, to sell joysticks, to sell light pens. So it was essential to seed the market.'

It also showed the software houses how serious Amstrad was about this new computer. 'We could go to the software houses and say, "Look, here's a computer, here's all the accessories, we're going to launch this and it will be in Dixons next September, we'll swear on the back of Bibles. And if you've got some games, we'll introduce you to Dixons and help you sell them in there. If you don't want all the hassle of selling to Dixons yourself, you can sell them to Amsoft and we'll publish them for you.'

Another crucial ingredient was building trust, and Perry didn't believe it was enough to deliver a computer, a manual, and an uplifting speech. 'If one of the software houses rang up and said, "We want to do something with this chip but we haven't got a data sheet for it' then I'd stick someone in a car and maybe around two hours later they'd arrive with it.' This even extended to hardware. Perry's team left each software house with the computer, a monitor, and a pile of tapes, making sure they had everything they needed. On those occasions when someone's monitor broke, an Amsoft employee was again deployed in a car to deliver a replacement in person. 'There was none of this, "Can't you just find one yourself, or I'll stick one in DHL and it'll be there by the end of this week."'

By April 1984, with the software finalised and the computer in low-volume production, it was time for the CPC 464's official launch. Amstrad, never afraid to outsource, had approached a public relations firm owned by Nick Hewer – eventually to earn fame as presenter of *Countdown* and one of Alan Sugar's original assistants on *The Apprentice*.

Hewer was to mastermind one of the most memorable computer launches ever, assembling hundreds of journalists in the Great Hall at Westminster School. As Hewer tells it in his autobiography, *My Alphabet: A Life from A to Z*, it took one of 'the two brainwaves I've ever had in my life' to come up with the idea [5]. Realising that the

CPC 464 has 'got colour, it's got sound, arithmetical function, music, and many other functions', and that it was aimed at children from the ages of eight to ten, what if he could find 'children with names that fit the functions?'

Together with a colleague, he unearthed a young Ravel, Monet, and Archimedes. 'We couldn't find a child called Shakespeare, so we had to make do with a 48-year-old William Shakespeare who was a worsted woollen merchant from Manchester.' It sounds like a recipe for the worst press event ever, but somehow Hewer and his team pulled it off – thanks in part to Alan Sugar taking to the stage and dishing out his 'pregnant calculator' line to dismiss the ZX81. Hewer describes the Amstrad share price as 'pretty much' doubling the following day.

The technical press were equally in love with the CPC 464, with Personal Computer World's Guy Kewney digging deep to pick out a criticism or two: 'It should have an indicator light on SHIFT LOCK and CAPS LOCK; and the serial interface is more important... than Amstrad realises' [6] were the best he could manage. More tellingly, he described Amstrad's all-in-one approach as a 'real marketing breakthrough' before going on to predict that the CPC 464 would 'comfortably outsell the Acorn Electron, and give the Commodore 64 and Sinclair Spectrum a hard run for their money'. And he even nailed a number to the post, predicting it would sell 200,000 units by the end of the year.

The CPC 464 even earned a positive review in the legendary American computer magazine Byte, with Dick Pountain lavishing particular praise on the operating system: 'it incorporates so many good ideas that I wished it had emerged four years ago when the Z80 was still hot' [7]. And while he clearly wasn't impressed by the now ageing Z80 processor, he described the ROM as being of 'fiendish ingenuity'.

This was back in late 1984. Interviewed for this book 35 years later, Pountain still remembers the CPC 464 fondly. 'Technically it was rather clever inside, not that anyone would ever notice,' he says. 'It achieved its performance at that price point by using very clever tricks. Somehow, Sugar managed to have a team of quite adept tech people working for him. Though not an equally good team of stylists.' For the sake of politeness, we will omit the exact word Pountain used to describe the CPC's physical design.

Jack Schofield was editor of The Guardian's Computing section at the time and remembered being equally won over. 'I thought the CPC was always a winner... it was cheap! It came with the monitor and the keyboard was quite good. And you didn't have to do anything.' This tapped into the so-called truck driver market. 'There was this huge untapped market of people that the Spectrum actually got into where you plugged it into a telly. But people actually wanted to watch telly, right? So the idea that you could give your kids something that included the monitor seemed to me a clear winner at the time.'

It's worth emphasising the price. At £229 for the green-on-black screen version, the CPC 464 was only £20 more than the ZX Spectrum+, which didn't include a monitor or tape drive and only came with 48kB of RAM to the 64kB that gave the CPC its name.

This perceived value, along with a colourful and bold design, meant the CPC 464 sold extremely well. Bearing in mind that the computer's initial run was for 100,000 units, it's telling that it went on to sell an estimated 2 million units in Europe. Judging from our own research for this book, we estimate the CPC 464 (and its later incarnations) sold around 1 million units in the UK.

So what was the CPC 464's impact upon Britain? It's true that it doesn't produce the same level of nostalgia as either the BBC Micro or the ZX Spectrum, but when we asked people via social media what impact the CPC 464 had on their lives we virtually drowned in responses.

For example, a young Jason Brown was begging his parents for a 'computer of my own, [but] they could only afford the green-screen 464. I didn't care, I had the world at my keyboard in my own bedroom. The world being 64K of memory and dodgy tapes from school mates. Fast-forward many, many years, and my love of computing, data and all things digital is still there and I'm chief digital officer of one of the largest companies in the UK (FTSE rank 39). I put it all down to the late hours and burning the midnight candle on my green-screen.'

We'll leave the final story to Luca Brumat. 'My mum bought me an Amstrad CPC 464 in 1986, when I was ten years old.' Having a keen interest in music, 'I immediately started to program my computer in BASIC 1.0 to play some simple electronic songs. I also chose an 'artist' name: Insert Coin. And year after year my songs became more and more complex. Thirty-three years later and I am an electronic music producer, I make records, I work with various labels. And the most important thing: my CPC 464 is still with me, fully working, and I continue to play games and chiptune music on it.'

The ARNOLD mystery

One of the problems Roland Perry faced when showing people the original CPC 464 box was that it had AMSTRAD plastered on the top-left corner. And Amstrad was extremely keen for people not to know it was behind this new computer. 'So we had to rub it off with nail varnish remover and Letraset 'Arnold' in the Amstrad house font,' recalls Perry.

So why Arnold? 'Later on, I swear, somebody said, "Oh, that's an anagram of Roland." And I thought, so it is, but I was too busy to notice at the time. The reason we put that, was so that we could go in there and pretend we were working for GEC.'

His logic was simple. By walking into a meeting and declaring that he was working for a 'large UK consumer electronics company', and then revealing a model of a computer with a great big Arnold sticker in the corner, people would put two and two together and think of GEC's chairman, Arnold Weinstock. And, according to Perry, that's precisely what happened.

Why CPC 464 rather than CPC 64?

Over the years, many people have wondered why Amstrad's first computer was called the CPC 464 rather than the CPC 64. To find out the answer for this book, Roland Perry went back to his original source: Bob Watkins. 'It was all about the product families,' says Perry.

Amstrad's convention was to use the first number to indicate which category each product fell into. So 1[000], 2[000], 3[000] and 5[000] were printers, while 4[memory] and 6[memory] denoted home micros. For PCs, Amstrad used 1[memory] (rather than 10) and then 2[memory], 3[memory]. 8[memory] were dot matrix word processors, 9[memory] were daisywheel word processors.

Mundane? Yes. But at least that's one more mystery solved.

What came next

The CPC 464 gave birth to five other official CPC computers, plus some clones, with the range finally being discontinued in 1990.

CPC 664

Release 1985   **Price** £339 mono screen, £449 colour screen

Introduced in May 1985, the 664's big innovation was an integrated 3-inch floppy disk drive (a decision we cover in the write-up of the PCW). Controversially, the CPC 664 was discontinued in late 1985 on the arrival of the CPC 6128, leading many CPC 664 buyers to feel short-changed.

CPC 6128

Release August 1985   **Price** £299 mono screen, £399 colour screen

As the name gives away, the CPC 6128 included 128kB of RAM, which meant it could support the CP/M 3.1 operating system. Amstrad initially released the CPC 6128 exclusively in the US (August 1985), for $699/$799 depending on the choice of screen, but it went on sale in the UK shortly afterwards.

464 Plus and 6128 Plus

Release September 1990   **Price: 464 Plus**, £299 mono screen, £329 colour screen; **6128 Plus**, £329 mono screen, £429 colour screen

Not only did Amstrad drop the 'CPC' precursor with this 1990 update, it also changed its focus to games – including a cartridge slot. Amstrad abandoned the built-in monitor, packaging all the electronics into a redesigned and rather dull grey keyboard, but both computers still shipped with a monitor that also delivered power to the computer unit. The 464 Plus included 64kB of memory and a cassette player, while the 6128 Plus featured a 3-inch floppy drive and 128kB of memory.

GX4000

Release September 1990   **Price** £100

In retrospect, this attempt by Amstrad to produce a video games console was doomed to failure. While it included two paddle controllers and a game – *Burnin' Rubber* – games developers had their eye on the worldwide release of the Sega Mega Drive and hotly trailed Super NES. Without games support, the GX4000 only sold 15,000 units.

Sources

Interviews with Rupert Goodwins, Mark-Eric Jones, Roland Perry, Dick Pountain, and Jack Schofield.

1. *Alan Sugar: What You See Is What You Get*, Pan Books 2011, page 253
2. As above, page 255
3. As above, page 253
4. As above, page 258
5. Nick Hewer, *My Alphabet: A Life from A to Z*, Kindle Edition (sample), location 130
6. Guy Kewney, Amstrad CPC 464 review, May 1984, page 170
7. Dick Pountain, *The Amstrad CPC 464*, Byte magazine, January 1985, page 401
    archive.org/stream/BYTE_Vol_10-01_1985-01_Through_The_Hourglass#page/n409/mode/2up

Sinclair QL

The British Macintosh that wasn't

There is a palpable sense of frustration when you speak to David Karlin, the project manager for Sinclair's ill-fated Sinclair QL. Announced one month before the Apple Macintosh, the QL was tantalisingly close to being a great British success: a low-cost rival to the Mac with a multitasking operating system, advanced processor, and striking design. But it wasn't to be.

'I wanted to do a few hundred pounds' worth of Xerox Star, bearing in mind that a Xerox Star would have been ten grand at the time,' remembers Karlin, who had seen an early version of that legendary machine during his time working for Fairchild Semiconductor in Palo Alto. Even in such a cut-down state, he believes, it would have been a machine 'capable of doing serious business'.

It was a compelling vision, and with Karlin's background in designing custom chips for Fairchild it's no surprise that Sir Clive Sinclair hired the young engineer at their first meeting. This was in August 1982, by which time Nigel Searle, the managing director of Sinclair Research, had already teased the UK press with its vision for the 'ZX-83'.

At first glance, Sinclair's vision – as set out by Searle – was similar to Karlin's. An affordable business computer? Check. A proper keyboard? Of course. Advanced processor and cutting-edge operating system? Absolutely. They even agreed that the QL should have an integrated display, although that key ingredient would be sacrificed late on.

Once you start digging into the details, though, a gap quickly develops. 'Clive's original concept for the QL was that it was going to be a super-lightweight version of the Osborne computer,' says Searle, referring to the Osborne 1, a 10 kg portable with a tiny 5-inch display and disk drives either side. Indeed, Sinclair's industrial designer, Rick Dickinson, built a prototype showing how this might work in practice as early as 1981 [1].

'Clive's rationale for the flat-screen television was not just as a television, but that it was going to be the display in the QL,' says Searle. 'So the QL was not going to require any peripherals, it was going to be a self-contained machine.'

By the time Karlin joined, however, it had become clear that this idea wasn't practical. 'The flat-screen TV certainly wasn't going to be ready in a QL kind of time frame,' he says. 'There was eventually a project to make a portable computer [the Pandora, which was followed by the Cambridge Z88] with one of those built in, but, to be honest, if the QL had its flaws, the flat-screen TV was a nightmare in manufacturing terms.' Even by 1985, by which time Karlin had taken charge of Sinclair's manufacturing division, it wasn't robust enough to work in volume.

But the real problem, and arguably the kiss of death for the QL, was Sir Clive's insistence on using the Microdrive for mass storage. After long delays, Sinclair had finally released the Microdrive for the Spectrum. It worked but wasn't the most reliable or quickest of storage devices, and there was also a communication problem: while the Spectrum exchanged data with Microdrives using analogue signals, the QL could only do so digitally.

This meant Karlin needed to use his chip-design skills to sample the stream of magnetic data being sent by the Microdrive and extract the relevant information. And before he did that, he needed to understand the duty cycle of the waveform: that is, the percentage of the wave that was 'zero' and how much of it was 'one'. If you think of it as turrets in a castle wall, then it's how much of the wall is peaks and how much lows. 50% is a perfectly even split, but if it's 10% then only one-tenth will be peaks.

Why does this matter? Because by knowing the size of the pulse widths, you know how often you need to sample the waveform to extract the right data. Otherwise, you might assume it's a zero when it could be a one.

This is the root of the ongoing problems with the Microdrive, and the reason why Karlin points to one moment in time that decided the fate of the QL. 'I would trace it to a specific conversation in a corridor with somebody called Ben Cheese, who's sadly died,' he says. 'Me to Ben: "What's the duty cycle of the waveform? You know, how far off 50% is it going to go?" Ben: "Oh, it won't go more than a couple of percent off 50." David goes away and thinks fine, that means I can sample it at, you know, four times the data rate, and I'll be fine… I don't need any analogue electronics and it'll decode the waveform properly.'

It's crucial to note, here, that Karlin isn't blaming Ben Cheese or any other individual for the QL's Microdrive struggles. If anything, this was a consequence of Sinclair Research not having safety catches in place. 'In a grown-up, sane organisation, a) I'd have been a bit more experienced and I would have had the nous not to just do that on the basis of a casual conversation in a corridor,' says Karlin. 'And b) there would have been an engineering manager somewhere who would have done some kind of design review and gone, hang on guys, this isn't going to work.'

In reality, the 'couple of percent' that Cheese suggested was more like 15 percent, which meant that – despite Karlin building in extra tolerance – when the Microdrive interface chips came back they didn't work. Or more accurately, says Karlin, 'they worked most of the time – but most of the time is ways off good. Ways off. It would have been better if it had worked a bit less because then it would have been categorically OK to delay this thing six months while we think of another way of doing it. But it actually worked well enough that it was possible to just about get things out the door.'

You might reasonably ask why, when faced with all these technical problems surrounding the Microdrive, Sinclair didn't move straight to the higher-capacity 3.5-inch floppy drives that were starting to appear. While expense is certainly one key reason – Sir Clive always had a tight focus on costs – another was pride. Let us not forget that Clive Sinclair was just as much an inventor as he was an entrepreneur, so if he and his team of engineers had created a new technology then he wanted to use it.

The processor was another radical departure, with Karlin persuading his boss that the cutting-edge Motorola 68008 was the right choice. By now, the Spectrum's 8-bit Z80 was showing its age, not just through its lack of raw speed but due to the fact it could only directly address 64kB of memory. While there were ways to dance around this, how much better for the QL to use a 32-bit processor?

Well, a theoretical 32-bit processor. Internally, Motorola's 68000 family of chips could indeed handle 32-bit data; that is, a piece of data consisting of 32 zeros and ones. However, to move data around the system – between the memory, the screen, and the disk, for example – you're dependent on the data bus. Which does exactly what its name describes. While Sinclair could have opted for a 68000 with a 16-bit bus, that would have incurred a huge extra cost. Knowing that he'd lose that argument, Karlin persuaded Sir Clive that they should use the lesser, but still powerful, 68008 chip with its 8-bit data bus.

Karlin had to make other sacrifices, too. For example, because integrated circuits cost '30p to 40p a throw, and we don't have lots of 30ps and 40ps to throw around', Karlin was told he could have two at most. Consequently, he had to rely on the cheap gate arrays available in 1983, which meant he had precisely 40 input/output pins to play with. 'I probably spent more time than anything else counting pins and going, well, is there a way I can multiplex these two functions on two pins instead of the three pins that you would normally think that they need?'

The end result for users was bottlenecks. One custom chip, which Karlin designed, would control the memory timing, send data to the display, and provide the communication link to the processor. The other would handle input and output to the modem, the Microdrives, the network, the keyboard, and the printer. This resulted in some major compromises that would haunt the QL.

The new processor ticked Clive Sinclair's desire to always be at the cutting edge, but meant that the company would have to create its own operating system. Rather than commission an external company, Sinclair turned to its in-house genius, Tony Tebby, who Karlin describes as 'the smartest person I've ever worked with'. And that's coming from someone who can design custom chips.

Jan Jones, who wrote SuperBASIC for the Sinclair QL (see 'The SuperBASIC story'), echoes this sentiment. 'Absolutely. The cleverest man I've ever worked with. The cleverest man I've ever known,' she says. 'His thinking was unbound really. I got on with him very well, because I could focus on the nuts and bolts of what he wanted – but I couldn't follow him into all the realms where his creative genius went.'

Things weren't so smooth when it came to relationships with senior management. 'Tony Tebby was undoubtedly a very, very talented programmer, but I found Tony a prickly person to get along with,' says Searle. 'And we had several fights, verbal jousts, over his absolute refusal to commit to any sort of deadlines. And we had all our eggs in one basket.'

With no certainty that the in-house operating system would arrive on time, Searle decided to move some of those eggs to a different basket and commission Cambridge's GST Computer Systems, run by Jeff Fenton, to write an alternative OS. 'My attitude was, you know, we've got to get this product out of the door,' says Searle. 'I'm just going to use whichever operating system is available and works.'

If you've ever watched a soap opera, you know what happens next. When Tony Tebby inevitably discovered that another company was working on an OS for the Sinclair QL, he was not best pleased. 'He sort of virtually went on strike at that point and stopped doing any work at all,' says Searle. 'Then he realised that if he didn't then he wouldn't have the glory of being the developer of the operating system for the QL, so he started working again.'

Of course, this is just Searle's viewpoint. Jan Jones remembers Tebby being 'justifiably cross' and then taking a few days off to make a point. 'He then cracked on to finish the job he'd started,' adds Jones. Tebby also deserves credit for how advanced QDOS was, with Karlin marking out the multitasking kernel for particular praise. 'It was a "proper" multitasker: you could spin up as many tasks as you liked and assign priority to each, allowing you to do things that wouldn't become possible on the PC or Mac operating systems for over a decade later – perhaps more.'

By this time, the ZX-83 codename for the Sinclair QL was starting to look questionable: there was no chance the QL would be released by the end of 1983. Some pessimists of the time might even have suggested that ZX-84 was hedging its bets. Nevertheless, and keenly aware that Apple was about to launch the Macintosh, Sir Clive committed to a launch event in January 1984.

This occurred with all the glitz and glamour that Clive Sinclair had become famous for. In a packed Inter-Continental Hotel, at London's Hyde Park Corner, he presented the wonders of the QL to the expectant press. Nigel Searle then stood up and walked the audience through the most important features.

The response was typified by Jack Schofield, The Guardian's Computing reporter, who, on the following Thursday, wrote [2]: 'Clive Sinclair last week launched his fourth new micro, the Sinclair QL. He said it was his most important introduction since the ZX80 revolutionised micro-computing in the UK. I think he was right. The QL fully lives up to its initials, which stand for "quantum leap."'

Speaking 35 years later, Schofield admitted he was taken in by the irresistible lure of Clive Sinclair's presentation skills and shiny promises. 'At the time, a multitasking operating system, 68000 processor, good software – it showed promise. I remember ordering one.' Then, after long delays, his unit arrived. 'It was rubbish,' said Schofield, without hesitation. 'I remember it being buggy as hell. And the early versions had a dongle at the back, didn't they?'

They did, and the reason was ROM. When Tebby's QDOS (not to be confused with the 'quick and dirty' QDOS Microsoft acquired from Seattle Computer Products; see the story of the IBM Personal Computer) and its accompanying

SuperBASIC interpreter landed, they couldn't fit on the allocated 32kB of ROM. So for Sinclair to start shipping, it needed to plug a memory expansion pack into one of the expansion ports: the famous 'kludge', which protruded from the back by a few centimetres. This was a temporary fix, so that the QL could finally start shipping, with Sinclair promising to replace these first-off-the-line QLs with a more complete unit when they started shipping in volume.

Clive Sinclair had previously promised that Sinclair QLs would start shipping within 28 days of the press launch, but anyone involved in the project knew this was not going to happen. Meanwhile, the positive coverage in the papers and specialist magazines meant that orders flooded in; according to *The Sinclair Story*, 'By the end of May, the company had received over £5m for 13,000 machines, but had only been able to deliver a few hundred.' [3]

The reason behind this slow progress was quite simple: the product wasn't ready. Those first few hundred machines were more like prototypes, with many bugs still to address. Not just related to the Microdrive either: early users complained about bugs in the Psion-provided software, which fell far short of the paradise Searle promised at the launch.

This software, like so much about the QL, sounded amazing on paper. In 1984, decent word-processing, spreadsheet, database, and graphics packages cost anything between £65 and £250 apiece. And you were getting all four of them for free, courtesy of Psion – a highly regarded software developer at the time, and already well-known for its Organiser pocket computer. While the software would eventually gain praise, and even an industry award, this was little consolation for early QL owners.

The first few months of 1984 saw Karlin fighting numerous hardware issues as he pushed the machine into full production, Tebby polishing QDOS (notably, the OS shown at the press launch was GST's, as this was closer to being finished at the time) and Psion working on the next revision of its software suite.

By mid-April, Sinclair Research finally had a working QL that it could hand over to David Tebbutt of computing magazine PCW. While acknowledging that the product wasn't yet finished, the review was surprisingly favourable, describing it as a 'well made piece of hardware' [4]. The final paragraph: 'The bottom line is that the QL gives you the potential to own a complete, serious computing facility, including printer and essential software, for under £1,000. Well under that if you're prepared to use a domestic TV rather than a monitor for the display.'

It was evident to Tebbutt, however, that this was not a business machine to rival the likes of IBM or Apple. It was hindered by its 100kB Microdrives, while even the high-quality Psion software was better suited to home use – the QL, he said, was 'not something to run a business on'.

Which brings us to the tricky topic of who, really, the QL was aimed at. Clive Sinclair actively did not want games software to be developed for the QL at launch – despite the presence of two joystick ports – as he wanted this to be a business machine. 'I think Clive really did want something that was more serious, grown-up, capable of bigger tasks,' says Karlin. Where he would have struggled, perhaps, is if directly asked who his target audience was and what, in Karlin's words, 'is really going to fire up the people who buy this thing?'

Karlin always had an audience in mind: 'People like my mum [who ran a small PR firm] and all her employees, and every office in the country. I don't know what Clive's answer would have been, but I don't think it would have been that. I don't think he knew people like my mum and I don't think he knew people who worked in typing pools. Or people who work in lawyers' offices where you need boilerplate and, you know, those kinds of things. I don't think that was his world.'

This conflict led to one of the design decisions that Karlin would most regret, with Sir Clive focused on producing the next great thing rather than really thinking about the business user. 'What I didn't get was a built-in monitor. And that was pretty crucifying because that was a decision taken very late on in the project. And it meant we had to bolt on a not very good domestic telly interface and that undoubtedly compromised the design.'

Ultimately, all these compromises, slow delivery, and a lack of third-party software – once again we can point to the Microdrives here, as only Sinclair could produce these and they remained slow and expensive – meant that the Sinclair QL's appeal was always limited compared to more focused competition. Businesses would be willing to spend the extra on an Apple Macintosh, IBM PC, or one of the rapidly growing number of IBM PC-compatible systems. Home users had the Commodore 64, Amstrad CPC 464, and the still-popular Spectrum to choose from.

Against this background, Sinclair Research was battling cash flow problems. It didn't help that it had always offered a generous returns policy, where people could return 'faulty' Spectrums, no questions asked, within a year of purchase. This led to Sinclair handing out thousands of replacements free of charge in 1985 against a background of slowing Spectrum sales, and ultimately led to the planned flotation of Sinclair Research being cancelled.

By this time, despite its early bugs and issues being solved, Sinclair believed that the only hope for the QL was to slash its price: by Christmas 1985, you could buy one for £199, half the launch price. This gave sales a temporary boost, but it wasn't enough to solve the company's cash flow problems. In April 1986, Amstrad bought Sinclair Research. Almost immediately, Alan Sugar quashed any plans for a second-iteration QL.

Despite this, the QL has a place in history: it was the QL that a young Linus Torvalds, creator of Linux, used to develop his coding skills. 'You had to play all these insane tricks,' said Torvalds in 2010 [5]. 'For example, the Microdrives were horrible. They were so speed-critical that you could not take the OS code and make your modifications and try to write to the Microdrives because now when you ran in RAM it was too slow.' He even wrote his own assembler and editor so that he could burn them to ROM and thus run them more quickly. Not quite the historic role that David Karlin and Clive Sinclair had sketched out for the QL back in August 1982, but not a terrible legacy for a computer burdened by so many challenges.

The SuperBASIC story

To call your programming language SuperBASIC is asking for trouble – what if it isn't Super at all? – but fortunately for its creator, Jan Jones, the QL's built-in BASIC interpreter lived up to its billing. Together with Tony Tebby, Jones was determined to create a more complete, grown-up version of BASIC. 'It was written by programmers for programmers,' says Jones, 'but with the added bonus that the ordinary user could also find it easy to run, easy to use as well.'

One of SuperBASIC's key strengths was that you could call on functions that you knew you were going to repeat, rather than typing them in every time. 'If you were building a piece of code, using one function you keep repeating, you could keep referring to that function; you didn't need to have it inside the code in every single subroutine, for example,' explains Jones. 'You had everything nestling at the bottom of the code. Everything had names. So you could just name it and the processor would go down and find it and bring it and do it.'

Her other big aim was to create something cohesive. 'Everything – all the instructions, all the commands – kind of looked the same,' Jones explains. 'Nothing looked as if it had been added on later as an afterthought or patch. We thought about every single element and how it would merge with the others. So I don't think that any of the features was exactly new, but the way we interpreted them and built them all together conceivably was.'

This integration extended to the QL, too, with SuperBASIC baked into QDOS (the QL's operating system) and burned to the ROM.

David Karlin believes SuperBASIC was on a par with 'anything around at the time', including C and Pascal, whilst 'performing all the functions of what is now called 'a shell script',' he says. 'To this day, I don't see why I have to write Linux scripts in a bastardised hack of a language knocked together in a hurry by UNIX's founders. SuperBASIC did it so much better.'

SuperBASIC has also survived the test of time. Consider that Jan Jones designed SuperBASIC in 1983/84, yet even 30 years later she was being asked for her original guidebook. 'I kept getting emails from people saying, oh, have you got any copies? Eventually I thought, well, I will just retype the whole thing, and I'll put it on the Kindle.'

It was a labour of love that took two months – the original text was long gone – but in November 1994 Jones uploaded it to Amazon where it continues to sell. Search for 'Jan Jones SuperBASIC' on Amazon to find the e-book, and you'll also find out what Jones now does for a living: her first work of fiction, *Stage by Stage*, won an award from the UK's Romantic Novelists' Association. A far cry from procedures and routines.

One Per Desk: the ICL spin-off

While ICL's One Per Desk doesn't look at all like a Sinclair QL – it's finished in dull beige, the dual Microdrive slots sit in a Toblerone-shaped tetrahedron above the keyboard, and there's the small matter of a phone handset built into the unit – but essentially it's the same machine. Lift up the lid, and the only difference is that ICL chose a more advanced Intel microcontroller for the telephony module.

The One Per Desk was aimed at executives, so rather than live with the ugly 2 kg power supply provided with the QL, this was integrated into the bundled monitor – a 9-inch mono or 14-inch colour screen. Sadly, much like the QL, they weren't a roaring success, with one contributor on **old-computers.com**, Ex Cathedra, describing it as a 'classic design disaster – great ideas, poor execution'. [6]

For instance, ICL improved the reliability of the Microdrives but removed the buffer, leading to long seek times. According to Ex Cathedra, it also suffered from overheating problems and 'frequent crashes'.

Sadly, we don't know how many OPDs were sold, but we do know that ICL handed Sinclair a timely £1 million investment during the QL's development.

When the keys dropped

If you ever wondered why the keys of your Sinclair QL kept dropping out, an interview with designer Rick Dickinson in Edge magazine from 2004 holds the answer. Recalling a visit to the factory in Italy where they made the moulds, he said: 'Now, the first time a mould is sampled, it's designed so you can adjust the parts. We pressed all the keys in, turned it upside down, and four or five fell out. We rattled it, and four or five more fell out.' [7]

Explaining that the mould needed to be fine-tuned to the production controller, Dickinson was told they didn't have time: doing so would delay shipping by a week. 'I was absolutely infuriated by the decision,' said Dickinson. But probably not as annoyed as all those QL owners who spent so much time picking up keys.

Sources
Interviews with Jan Jones, David Karlin, Jack Schofield, and Nigel Searle.

1. Rick Dickinson, QL and beyond album, Flickr, uploaded 8 January 2015
   flickr.com/photos/9574086@N02/16205461566/in/album-72157600854938578
2. Jack Schofield, *Keeping one leap ahead*, The Guardian, 19 January 1984, page 15
3. Rodney Dale, *The Sinclair Story*, Gerald Duckworth & Co Ltd 1985, page 138
4. David Tebbutt, Sinclair QL review, Personal Computer World, June 1984, page 170
5. Jeremy Allison, *Geek Time with Linus Torvalds*, Google Open Source Blog, 27 September 2010,
   opensource.googleblog.com/2010/09/geek-time-with-linus-torvalds.html
6. ICL One Per Desk, Old-Computers.com, undated
   old-computers.com/museum/computer.asp?c=275&st=1
7. *The Rick Dickinson guide to classic computer design*, Edge magazine, issue 137, June 2004, page 81

Atari 520ST

**A gaming phoenix rises
from the ashes**

That the Atari ST series exists at all is something of a miracle. That it was so good wanders into the realms of 'how on earth did they build pyramids in 2500BC?' territory.

Consider this: in 1983, Atari posted losses of $538.6 million based on a $1.12 billion turnover. With the video console market crashing, the company's revenues had halved in the space of a year. Little wonder that Atari's owner, Warner Communications, was looking to offload the company in early 1984. Its CEO, Steve Ross, figured he had the answer in Jack Tramiel.

The ex-president of Commodore had only resigned from the computing giant in January. While Tramiel's stated plan was to travel the world with his wife Helen for a year, it only took a phone call from Ross to lure him back to California. This was in April. Within days, Jack had created a new company called Tramel Technology Ltd with the aim of building a next-generation computer, either through Atari or on his own. (For trivia fans, there's no 'I' because Jack was fed up of people mispronouncing his name.)

We asked his son, Leonard, whether Jack was hooked on the idea of buying Atari when he heard it was for sale. 'No, he was *interested* from the moment he heard they were for sale. Atari was the second best-known brand in the world after Coca-Cola,' he says. 'And having the running start of an infrastructure was an obvious plus. But it was only because he was able to put together a fantastic deal that he took it.'

After three months of negotiations, during which time Atari's agreement to build a computer using Amiga technology controversially fell through (see the Commodore Amiga story), Tramiel took ownership of Atari's home video game and computer businesses on 2 July 1984. While no cash exchanged hands, Atari needed a cash injection of $30 million from Jack and others to keep it going. Not a bad price for what was then the second best-known brand in the world.

Such was Tramiel's reputation among the senior Commodore team that many had more loyalty to him than the billion-dollar company he had helped to create. In May, against a background of his ongoing Atari negotiations, he was meeting with Shiraz Shivji – who had been director of R&D at Commodore – to discuss what that next computer could be. Legend has it that Shivji even sketched out what essentially became the Atari 520ST during one of those meetings.

At the point of sale, Atari had around 900 employees. They all now had to reapply for a much smaller number of jobs, with around 300 retained, while Tramiel also hired around 30 trusted and talented engineers from Commodore. Including Shivji. Commodore was not impressed at the brain drain: the company quickly sued Shivji and three others for stealing trade secrets, while Tramiel later countersued for $50 million over the aborted Amiga deal.

But there was no time to be distracted by legal battles: Shivji had a computer to build. Tramiel wanted Atari to make a statement at the Consumer Electronics Show in January 1985 by debuting its new computer, and that meant making a lot of decisions and fast.

The first decision was which processor to build it around. 'We were hot on the [32-bit] 32016 and 32032,' said Shivji in 1988 [1]. 'We had a bunch of meetings with National Semiconductor regarding the availability of the chip, and when it was obvious that we could not have the number of chips that we wanted and the pricing was not right, then the decision was made to go with the 68000.'

In an interview for Imagine Publishing's *Classic Videogame Hardware Genius Guide* from 2011, Shivji explained that, right from the start, 'music and graphics were already important for us' [2]. Initially the Atari engineers, including four from Commodore, had hoped to use an advanced audio chip custom-made by Atari, but this proved bug-ridden and expensive. They compromised by choosing an off-the-shelf Yamaha chip, but complemented it with MIDI-in and MIDI-out ports. A decision that would make the Atari ST a much-loved choice among musicians.

This was also an expandable machine. You could add a ROM pack via an expansion slot on the left-hand side, while the right-hand side offered two joystick ports (although one was normally occupied by the two-button mouse). And keeping the MIDI ports company at the rear, take your choice from (deep breath) an RS-232 serial port, Centronics parallel port, TV video out, RGB video out, composite video out, floppy disk connector, and hard disk connector.

Then there was the case itself. Designed by Ira Velinsky, yet another Commodorian, it splices the DNA of the Atari 400 and 800 computers with the Commodore Plus/4, another of his previous creations. While it was a machine where form followed function, it had enough style – note the slanted function buttons, for instance – to give it a futuristic look.

What's perhaps most impressive is that, in the space of six months, Atari developed four custom chips to support the Motorola 68000 processor. Thanks to the custom-made video chip, for instance, the 520ST could output at a 640×400 resolution in black-and-white. Switching to 320×200 stretched that to 16 on-screen colours out of 512 colours in its palette. Sure, the Amiga could do even more, but that was also twice as expensive; it's little wonder that Shiraz stated that he 'was very proud that the team accomplished so much in a short period of time'.

Shiraz's team wasn't alone in working ridiculously hard to ensure Atari had a machine to show at the January 1985 CES. Atari knew that its new computer needed a graphical user interface similar to the Macintosh – the Atari ST quickly earned the nickname of a Jackintosh – and effectively had two choices of Digital Research's GEM (Graphics Environment Manager) and Microsoft's upcoming Windows 1.0. 'We went to DRI and Microsoft,' says Leonard Tramiel, who was in charge of software development at the new Atari. 'And Microsoft said that in order to meet the time deadline we had, which was insane, they could not port to our machine unless it was an Intel-based architecture.'

Fortunately, Digital Research was a lot more open to working with Atari on the project, so Leonard moved his team of a dozen or so software engineers to the company's base in Monterey, California. Effectively, they had six months to port GEM from Intel's architecture to the Motorola 68000. 'It was just an enormous amount of work, to say nothing of all of the graphics routines that used hardware that was different than anything else that had been done,' says Leonard. 'I'm not sure I can put it in a way that makes it obvious to someone not versed in the field, but it was a hell of a lot of work.'

It didn't help that Digital Research hadn't actually settled on final code for GEM at this time. 'The killer was that Digital Research would give me this 8086 code and say, "Translate this",' said Dave Staugus, one of Tramiel's team of programmers [3], 'and then a week later they'd say, "Oh, there was a lot of bugs in that, here's the new one."' And while all this was happening, Leonard had to oversee the development of TOS – with the T standing for The, Tramiel or Total, depending on who you speak to – as the disk operating system.

But, between the hardware and software teams, Atari delivered working prototypes for CES, as planned. Albeit by the narrowest of margins. '[I remember] somebody telling me that we were waiting for a couple of our engineers to come from the chip fab because the first prototypes of two of the custom chips were coming,' says Neil Harris, a former 'VIC Commando' who had jumped ship to Atari to run its magazine division, 'and nobody knew if they were actually going to work or not because they'd never been tested. They had to be plugged into the computer and turned it on. And suddenly it worked. That was a miracle.'

The Atari ST was one of the hits of the show. 'People were stunned,' remembers Leonard Tramiel. 'One of the tech reporters flat out refused to believe that a colour machine with that level of performance had been built with the available technology. He was convinced that we were showing a mock-up that was just acting as a video display for a mini computer that was in one of the back rooms of the booth. So to prove to him that isn't what was going on, we found an electrical outlet in the middle of the floor, plugged a light bulb into it, so he knew that it was just delivering power, and plugged our machine into it. And until he saw that he did not believe it was real.'

Atari delivered even more buzz thanks to its price. It announced the Atari 130ST, which was to include 128kB of RAM but never shipped, for $399, and the 512kB 520ST for $599. With no sign of the Commodore Amiga at CES that year, and no other big announcements of note, it was an undoubted triumph for Jack Tramiel. Atari was even promoting a slew of peripherals to support the ST, from 3.5-inch external floppy drives to colour printers, to a choice of mono or colour monitors.

There's a big difference between delivering prototypes at a show and shipping computers to high-street retailers, however, so it took several more months of hard work to complete both the hardware and the software. And all the while, there was the lingering threat that the $30 million Tramiel and partners had invested would run out before the 520ST could go on sale: Jack Tramiel described sales of Atari's existing 8-bit product line as 'very, very slow' [4] in early 1985.

Question marks also hung over distribution in the United States, with one retailer saying their interest in doing business with Tramiel's Atari was 'zero, zilch' [5]. This was due to Jack Tramiel's past history at Commodore, where retailers were sometimes left with stock that they had to sell for a lower price than at which they had bought it after Commodore announced a price reduction.

Another annoyance: Atari's disk operating system TOS wasn't finished when the first 520STs were shipped to retailers in June 1985. That meant users had to load TOS from a floppy drive rather than from ROM. 'This greatly reduced the amount of usable memory,' wrote an early 520ST user [6]. 'I still remember the disappointment of booting up my brand new half-megabyte ST, loading in ST BASIC and finding that I had only seven kilobytes of RAM left for code!' Fortunately, Atari made good on its promise that it would retro-fit the ROMs when TOS was complete, which happened later in 1985.

The biggest problem for new Atari 520ST buyers was the same that afflicted virtually every new platform: a lack of things to do with it. While Atari was eager to point out that the Motorola 68000 was the same chip used in the Macintosh, and that it would be easy to port software from the Mac to the Atari ST, there were still significant differences between the two machines. This process would never be trivial.

It was an issue that would also dampen the Amiga's appeal when it was properly released in early 1986, but it's worth noting that Your Computer Magazine was complaining about the lack of software for the Atari even in April 1986. 'The only real criticisms invited by the new machines [the 520STM, 520STFM, and 1040STF] is the software base. Standing at slightly more than 100 titles, the ST range pales against the 5,000 or so available to users of the Sinclair Spectrum.' [7]

One section of the audience that immediately appreciated the Atari, though, was programmers. Previously, a Motorola 68000 machine cost significantly more than £1,000, but thanks to the efforts of Metacomco's Tim King (who also wrote Amiga DOS) early users could benefit from a suite of development tools. 'We did one in Assembler, we had a Pascal compiler, we had a C compiler, we had a BPCL compiler of course, we even had a Lisp system,' says King.

British Atari ST user – and fan – Simon Hudson agrees. 'I wrote my first useful code, and started to really understand how technology could enable business change [on an Atari ST],' he tells us. And Simon went on from there to start a digital workplace and analytics company that 'created the world's first out-of-the-box SharePoint enterprise intranet'.

Press reaction to the 520ST was a cautious welcome. 'There can be no doubt that the 520ST is a very impressive machine,' wrote Peter Bright in Personal

Computer World [8], praising the keyboard, expandability, the 512kB of RAM and 'what, for many, is the nicest 16-bit processor around'. But he also cautioned people to consider the lack of software and the total price: '...you are looking at about £700 for a 520ST and disk drive. When you add a monitor and other bits and pieces, you won't see much change out of £1,000.'

Even when you added up the extras, though, the 520ST offered excellent value for such an advanced computer. Especially when compared to the first Amiga, which cost almost £1,700 including VAT. This, combined with the delay to the Amiga, gave the Atari an excellent head start over its rival and helped it to sell a respectable 100,000 units in its first year.

This, along with its advanced graphics capabilities (and the fact it was relatively easy to create a game for the Atari and then port it to the Amiga), encouraged games developers to focus their efforts on the new platform. Even by 1986, Atari ST buyers could enjoy Microdeal's *Time Bandit*, with *Thrust* clone *Oids* following in 1987. And *Xenon 2* remains a classic.

Despite these forward pushes, and the 1986 release of the 1040STF with 1MB of memory and a built-in floppy disk, the Atari ST never quite built enough momentum to fully deliver on its undoubted promise. It was actually far more successful in Europe than the US, but that's always been the wrong way round if you want to be a worldwide player. Ultimately, the figures don't add up unless you dominate the US market too.

So why did the Atari ST succeed more in the UK and Germany than in its home country? ' "People never got fired for buying IBM" had apparently never been translated into German,' says Leonard Tramiel dryly when we put this question to him, 'which is why the ST was such a success in Germany. And then Apple had marketed themselves as a very niche artistic machine and we couldn't touch that. I regret to this day not pushing harder on our advertising folk to make a commercial shortly after the ST came out, showing someone using a mouse and drawing an Apple logo, which at the time had these coloured stripes in it. And point out that the only machine you could do this on was an Atari ST.'

It didn't help Atari's cause when Commodore finally figured out that it should split the Amiga A1000 into two, with the A500 for the home market and A2000 for professionals. With an Amiga A500's more advanced graphics now selling within touching distance of the ST's price, the Atari began to look isolated and dated.

As Neil Harris points out, the Atari ST was also a victim of its time. 'In the early days [of a market], you're making it up as you go along, it's much more about being a cowboy. And later on, as the market starts to mature, standardisation sets in; it becomes less fun but much more successful and bigger. And that's kind of what happened to us.'

What does the ST stand for?

Never let techies name computers. 'ST stands for Sixteen/Thirty-two (16/32) – the Motorola 68000 is a 32-bit processor and communicates through a 16-bit bus,' explained Andrew Reese, editor of STart magazine in December 1989 [9]. 'Motorola's newer 68030 processor is also a 32-bit processor but it communicates through a 32-bit bus.' And that's why the ill-fated Atari TT earned its name.

Atari's 8-bit computers

Way back in 1979, several years before the 520ST went on sale, Atari made its first foray into home computers with the Atari 400 and 800. They sold well in the US, but by the time they found their way to UK shores in 1981 they looked expensive compared to the £140 Acorn Atom and £70 Sinclair ZX81: even a generous 16kB of RAM couldn't soften the 400's £395 price, while the Atari 800 cost £695.

These were, though, extremely well-engineered machines that could take far more of a beating than their British-made equivalents. Add strong documentation and high-quality peripherals (such as a 90kB disk drive and 80-column printer) and they had much in their favour. Enough for Personal Computer World to declare that 'they might spawn a completely new "consumer education" boom' [10] in its October 1980 preview.

While that consumer education boom certainly happened, Atari's 8-bit computers only played a bit part in the UK. If they had been sold in Europe earlier, even in 1980, it might have been a very different story – and that might also have eased the way for the Atari ST range almost a decade later.

What came next

The Atari 520ST spawned a family of computers that kept the Atari ST brand alive until 1993.

Atari 1040STF

Release 1986   **Price** £799 with mono monitor, £899 with colour monitor

In its 1986 refresh, the 1040STF was the prize model – although to avoid disappointment when playing games it made sense to buy the colour version. A redesigned chassis integrated a floppy drive (the 'F' in the name) and included 1MB of RAM.

Atari 520STFM

Release 1986   **Price** £499

A cut-down version of the 1040STF, with 512kB of RAM and no bundled monitor. Instead, Atari included a RF modulator (the 'M' in the name) so you could output to a TV.

Atari 520STM

Release 1986   **Price** £399

The last of the 1986 updates, and decidedly the least, the 520STM used the redesigned chassis but lost the floppy drive.

Atari ST Mega

Release 1987   **Price** £999.95

Squeaking under the £1,000 mark, the Atari ST Mega targeted businesses with a built-in 720kB floppy drive, monochrome monitor, and mouse. The £999.95 version included 2MB of RAM, but you could double that to 4MB for £1,299.95.

Atari 520STE/1040STE

Release 1990   **Price** £359/£439

With the 'E' standing for enhanced, the STE looked similar to its predecessors but promised 4,096 colours – although only 16 could be displayed on-screen at the same time. You could also attach stereo speakers.

Atari TT

Release 1990   **Price** £2,350

Even by Atari standards, the TT was beset by delays; it was announced in 1987. Much of the delay was due to battles with Motorola's 32-bit 68030 processor, but when it did arrive there was support for up to 26MB of RAM and a new version of the GEM Desktop – complete with the ability to drag files and folders from windows onto the desktop.

Atari Falcon030

Release 1992   **Price** £499

Against a backdrop of quarterly losses, Atari pinned much hope on its advanced Falcon. This included a multitasking version of TOS, a Motorola 68030 processor, up to 65,536 colours on-screen and support for up to 14MB of RAM. The £499 price is for the 1MB version: for £899, you could buy one with 4MB of memory and a 65MB hard drive. Sadly, the Falcon never flew off the shelves, and was discontinued in 1993.

Sources
Interviews with Neil Harris, Tim King, and Leonard Tramiel.

1. Jeffrey Daniels, 3 Years With the ST, STart magazine, Summer 1988, page 22
    atarimagazines.com/startv3n1/threeyearsofst.html
2. Chapter 14: Atari ST, Classic Videogame Hardware Genius Guide, Imagine Publishing, May 2011
3. Jeffrey Daniels, 3 Years With the ST, STart magazine, Summer 1988, page 22
    atarimagazines.com/startv3n1/threeyearsofst.html
4. Tom Maremaa, Atari Ships New 520 ST, InfoWorld, 3 June 1985, page 23
    books.google.co.uk/books?id=8C4EAAAAMBAJ&q=maremaa#v=snippet&q=maremaa&f=false

5. As above

6. William Hern, response to 'The 68000 Wars, Part 2: Jack is Back!', 2 April 2015
    filfre.net/2015/04/the-68000-wars-part-2-jack-is-back

7. Geof Wheelwright, Atari 1040STF - a powerful performer, Your Computer Magazine, April 1986, page 51

8. Peter Bright, Benchtest: Atari 520ST, Personal Computer World, June 1985, page 136

9. Andrew Reese, The Future of Atari Computing, STart magazine, December 1989, page

10. David Tebbutt, Benchtest: Atari 400 & 800, Personal Computer World, October 1980, page 63

Commodore
Amiga

**The PC beater that never
quite beat the PC**

A bouncing ball, a flight simulator, and a penny-pinching board of directors. These were the key ingredients that gave birth to the legendary Amiga series of computers.

Without the Atari board cheating its engineers and programmers out of their bonus, 'father of the Amiga' Jay Miner might never have left the company. If Miner hadn't seen a flight simulator in action, he might not have decided that he wanted to build a low-cost computer with the Amiga's ambitious graphical capabilities. And without a bouncing red-and-white ball, the whole Amiga project might have been stamped out of existence at prototype stage.

The Amiga's foundations can be traced back to 1979, when Atari was at its peak: not only was the 2600 console shipping by the million, but its new computers – the Atari 400 and 800 – had captured the imagination of the American public. Time to reward all those talented people who had created the machines? In what is surely one of the most short-sighted decisions in history, the Atari board decided it would say a very special thank-you by instead withholding the bonuses they had promised.

It should be no surprise that all Atari's key talent left. Larry Kaplan, who was chief programmer, set up a new games company called Activision along with several other Atari developers. Jay Miner considered staying, but when his bosses told him that they didn't want to create a new computer based on the Motorola 68000 – the obvious next move, in his view – he quit too, joining a chip manufacturer called Zimast. This made custom pacemaker chips for a company that would become integral to the early Amiga: Intermedics.

Three years passed, during which Activision produced hit after hit for the Atari VCS, but then Kaplan hatched a plan. 'Doug Neubauer and I went to talk to Jay Miner and the company he worked at about doing a new game system,' said Kaplan [1]. 'I had seen the Nintendo NES at the CES in June '82 and thought we could do better. The president of [Miner's] company contacted the owner in Texas and by October we had hired a president from Tonka Toys and got $6 million in funding.'

The idea was for Jay Miner to stay put at Zimast but still design the chips for the new video console, with Kaplan taking charge of game design. Unfortunately, Kaplan called the Atari president Nolan Bushnell to see if he wanted to be chairman of the board. 'When I called Nolan, he said I could do better financially with him, so I quit what was to become Amiga.' With Kaplan gone, Neubauer also abandoned ship.

This left Dave Morse, their high-profile Tonka Toys recruit, with 'the offices, a business plan, and financing,' explained Miner [2]. 'The financial backers still wanted a video game company, so Dave Morse asked me to take Larry Kaplan's place. This meant leaving Zimast. Dave Morse was president and I was vice-president.'

At last, Miner could realise his dream of building a super-fast computer based around the Motorola 68000. Except that Intermedics – the consortium backing the company headed by Texan billionaire Wayne Rollins – had signed up to create a video console, not a computer. In a stroke of genius, Miner decided to give them the console they wanted but design it in such a way that it could be the basis of a computer as well.

He also had a simple ambition that would help guide his design decisions. 'My goal was to design a low-cost computer that could do good flying aeroplane simulations,' said Miner. 'My friend at Singer Link, Al Pound, had shown me the real million-dollar simulators and I was hooked. I had to have a low-cost version of that to practice on at home.'

Now to build the team. One of Miner's first recruits was Bob Pariseau. 'They had the idea, but you can't sell a computer without software, so they hired me to put together a team to make the software. That was May of 1983,' he said [3]. 'We had seed money to do the software, but not much sleep… One of the guys used to walk around the room with a pillow in his arms, so he could sleep while he walked.'

They were under pressure because they knew they had to make a splash at CES, the big electronics show in Las Vegas, the following January. However, Miner and Morse didn't want the rest of Silicon Valley to know what they were up to, so as a distraction – and to bring in some cash – they started selling joysticks and other gaming hardware. One such product was The Joyboard: this balancing board, much like the Nintendo Wii's Balance Board, could hook up to an Atari 2600 console and then control what happened on-screen. One of the Amiga's key programmers, RJ Mical, developed a mini game called *Zen Meditation* to help him and others relax: you placed The Joyboard on a chair and then had to sit perfectly still. That's why early versions of the Amiga operating system included the message 'Guru Meditation Error' when something went wrong.

They also benefited from a cunning deflection technique in case people overheard them talking about the three chips they were developing. Denise was the main video processor and also supported eight sprites (useful for fast-moving games). Paula's main duty was to handle sound, including the digital samples that games developers would come to love. And Agnus hooked directly into the memory, while also containing two special graphics chips of her own: a blitter and 'Copper'.

The blitter could quickly copy and then redraw existing elements in the video memory without the aid of the main processor; Miner wanted this as a 'low-cost way to improve animation, such as flight simulators'. Copper, an evolution of the Atari 400's ANTIC chip, proved enormously useful to programmers too. This coprocessor could again act independently from the main CPU and enhanced its video capabilities in several ways. For instance, it allowed the Amiga to show three windows each with different resolutions and colour depths. And if you ever saw a game or demo use a 'raster bar' effect, you had Copper to thank.

To complete the female entourage, they named the assembled machine Lorraine. All those involved with Lorraine's early development paint a picture of a team working hard and in harmony. Miner and Pariseau didn't rule with an iron first, instead listening to what their engineers proposed; Pariseau's software team literally hammered out decisions with one of the foam baseball bats that he bought to lighten the sometimes intense mood.

Then, in mid-1983, disaster struck. 'So there we were designing this super graphic computer with four blitter channels, eight sprites, and four sound channels and the bottom just fell out of the video game market,' said Miner. 'This killed the joystick half of the company, and the cartridge market and that half of Amiga started losing money fast.'

Suddenly, Miner's Plan B to build a powerful but low-cost computer became Plan A. Amiga's investors backed this decision but had problems of their own; in late 1983, Intermedics stopped funding Amiga, which now had to pay its own bills. The engineers continued to work but, explained Miner, with 'severe financial restrictions. It seemed like we owed money to every supplier in town. I had to mortgage practically everything I owned personally to help meet the company payroll.'

Amiga president Dave Morse and his business advisor Bill Hart took the Amiga concept to a series of investors, but with so much uncertainty around the future of computers, video consoles, and the total cost of the project, no one bit. So they gambled and approached Atari. They revealed the technical details of the chips they were developing, which would fit in perfectly with a next-generation video console. A console that Atari now needed due to the rapid fallaway of its 2600. Morse's idea was for Atari to buy the rights to use the chips in a console while Amiga would continue to work on its computer.

Fortunately, Atari was interested. In late November 1983, with just six weeks until CES 1984, it signed a letter of intent prior to a full licence negotiation. At this point, though, no money exchanged hands. Morse and Hart also had another big problem to solve: 100% of Amiga was owned by Intermedics. Bill Hart had managed to raise around $1 million in pledges to support Amiga from smaller backers, but without equity they had nothing to offer in return.

Amiga had only one thing in its favour: time. Intermedics' tax year ended on 31 December, at which point it would have to write off Amiga as a multimillion-dollar loss. Dave Morse and Bill Hart flew to Louisiana on 30 December to negotiate with Intermedics' attorneys the following day. 'The lawyers had been charged with negotiating this and we didn't even have a lawyer with us,' said Hart [4]. 'So we sat down with them and their idea of how to negotiate with us was fairly strong armed… These guys would yell at us. "If you think you are going to steal this company out from under us, you are mistaken!"'

The two men stayed calm under verbal fire, as they knew the midnight deadline was ticking closer. After many wasted hours, they finally reached an agreement: Intermedics would be given 25% equity of the company in return for 'forgiveness of the debt'. This gave Amiga equity to share among its loyal employees and investors.

And Amiga still desperately needed investment. With CES just days away, the prototype was nowhere near being a production machine; Miner's three chips only existed as 'eight or ten circuit boards, stacked together, all interwired,' according to Pariseau [5]. Each of those stacks got its own seat on the flight to Vegas, with one famous photo showing a stack with a jacket around it and a pillow with a drawn-on smiling face. The two Amiga engineers sitting between it tried and failed to persuade the airline staff to give their new friend a meal.

They needed the sustenance, too, as it was going to be a long night. 'At the start of the show the next morning,' remembered Pariseau, 'we came in and saw the engineers were asleep on the floor inside the booth. They had stayed up all night to make another demo. It was the bouncing soccer ball. They had done it overnight.'

While it doesn't look anything special today, the Amiga's bouncing ball was revolutionary for 1984. Consisting of red and white checks, the large three-dimensional ball bounced up and down, obeying physics in a way that simply hadn't been seen before (except, perhaps, on a computer costing tens of thousands of dollars).

Amiga's space on the show floor was a booth with its hardware gadgets on show, and a room tucked away at the back. Visitors ranged from investors to press to department stores. 'We got a buyer from Sears in there,' said Pariseau, 'we got to the point in the demo where we showed the bouncing ball, and this guy stood up, pointed at the screen and said, "That cannot be done!"'

It was a hit with press attendees too. Creative Computing magazine even declared the Amiga prototype its 'hit of the show'. 'Suffice for now to say it is the most amazing graphics and sound machine that will ever have been offered to the consumer market,' wrote John Anderson [6]. 'Lorraine is capable of providing multi-color real-time animated images on a par with (and probably superior to) Saturday morning cartoons.'

Thanks to the showstopping demos, excellent press coverage, and his own selling powers, Bill Hart closed the deal with his multitude of investors, bringing in enough money for a few more months. However, this wasn't enough time to finish the Amiga, which meant the Atari deal was still crucial. The two companies finally agreed that Atari would pay a $2 royalty for every console it sold that contained the Amiga chipset, rising to a $15 royalty if it was included in a coin-op machine.

In search of cash up front, Hart suggested they buy shares, but Atari counter-offered. It would loan Amiga $500,000 for signing a letter of intent, with a further $1 million if and when the two companies signed the final licensing agreement and $500,000 for each chip. That brought the total to $3 million. In effect, this would be a loan to ensure production, and when the chips were finished Amiga would pay it back in the form of a million shares at $3 per share.

By the time the two companies signed the letter of intent, on 7 March 1984, a few crucial clauses had been added. Two were to Amiga's distaste: first, that as of 1 June 1985, Atari could extend its video console with a keyboard and disk drive, pushing it towards computer territory; second, from March 1986, Atari had the rights to build a personal computer of its own based on the Amiga chips. But the letter of intent also stated that if Amiga paid back the $500,000 loan in full by 30 June 1984, then the deal was off and it was free to go its own way. If, on the other hand, Amiga didn't pay back the loan and failed to sign the full licensing agreement, Atari would own all of Amiga's assets and intellectual property.

With cash in its pocket, Amiga pressed go on chip fabrication and the following month the three chips arrived in the office. To take a design from sketches to silicon in just over a year was incredibly quick; while there were bugs to fix, it meant that by the time June 1984's CES rolled around Amiga had fully working prototypes to show on the main floor. This was good news for Morse, because it greatly increased his chances of finding a new buyer so that he could pay back the Atari loan and cancel the agreement.

Demos once again wowed the crowds. The bouncing ball demo, entitled 'Boing', had been enhanced with sound: 'the booming noise of the ball was Bob Pariseau hitting a foam baseball bat against our garage door,' explained Miner in Brian Bagnall's book, *Commodore: The Amiga Years* [7]. It now bounced side to side too, using the edges of the monitor as virtual walls. A public used to computers that could only display a handful of colours at a time were equally amazed by a rainbow demo, which took advantage of all 4,096 colours Lorraine was capable of producing thanks to its HAM (hold-and-modify) mode.

With the addition of real-time multitasking, many people simply didn't believe Lorraine was doing the heavy lifting. 'It was wonderfully gratifying,' said Mical [8], 'because the more savvy people invariably walked up after the demo and gave a good look at the machine. They would get down on one knee to lift up the skirt and look under the table to see where the real computer was.'

There was plenty more to wow the crowds. To show off its audio and speech-synthesis abilities, engineers would type in a sentence for an on-screen robot to say. 'We would ask the audience to throw sentences at us, and we would type them in and it would speak those sentences for you right there,' said Mical. 'When one of the bigwigs at Sears was in there seeing the demo, without warning, the guy who was riding the keyboard typed in, "I buy all my tools at Sears", and the place lit up in a great roar of laughter. It was a wonderful little touch.'

By now, Amiga was confident enough in the final product to distribute spec sheets to press, software houses, and peripheral makers. The spec included a coprocessor slot (this would allow people to buy an add-on card and run MS-DOS, if they wished), a 3.5-inch floppy drive, 128kB of RAM, and a mouse. Morse promised a price of under $2,000 and, rather more ambitiously, that it would be ready for Christmas 1984.

Before then, though, there was the small matter of escaping the Atari deal. Negotiations between the two companies hadn't been going well, so Morse's great hope was that Commodore would swoop in, repay the money to Atari, and then create the Amiga home computer.

There were two things in his favour. First, Jack Tramiel – the man who brought Commodore to its current heights, but had left in early 1984 after a disagreement with its chief financier Irving Gould (see the Commodore 64 story) – was in the process of acquiring Atari. Nothing would please Gould more than snatching Amiga from Tramiel's clutches. And second, Commodore needed a follow-up to the C64, but had no suitable technology in the pipeline and had also lost a clutch of its best talent to Tramiel's new venture.

In the middle of June, Commodore sent a squadron of engineers to California to take a closer look at the Amiga technology. 'I went out there in this big rush because we have to do this before Atari take possession by default,' said Bob Russell [9]. 'I took a bunch of chip designers, software guys, and hardware guys out there.'

Russell's report would make joyful reading for Amiga, declaring 'they've got exactly what we need as far as a chipset core and features'. By the end of the month, Morse and Miner were holding a fresh letter of intent: Commodore had granted Amiga a $750,000 loan to cover the $500,000 it owed Atari and for ongoing expenses. This allowed Morse to return the money to Atari.

Although even this had been tricky. Atari had already started planning a new console based on Amiga technology, so to have it stolen away wasn't welcome news. At first, Atari's legal counsel had refused to accept the cheque, only for Tramiel – who happened to be in the building negotiating his takeover of Atari – to utter yet another memorable phrase. Bill Hart recalled the scene for Bagnall's book [10]: 'Jack said, "When somebody hands you a cheque for $500,000 you take it. You get it to the bank quickly."'

At that point, though, Jack Tramiel wasn't privy to all the background. When his son Leonard came across the legal agreements and discovered that Amiga had wriggled its way out of a deal – which included the promise that it wouldn't negotiate with other companies – Atari launched a $50 million lawsuit [11]. This dispute would rumble on for three years, along with a counter-suit from Commodore against Atari for stealing its engineers. The dispute finally ended in March 1987 when Commodore agreed to pay Atari an undisclosed sum – rumours vary from less than a million dollars to $8 million – on the day the trial was due to start.

But that was in the future. For now, Morse needed to squeeze the best deal possible from Commodore, and once again he proved a shrewd negotiator: Commodore eventually bought Amiga for a generous $23 million. It created a new subsidiary called Commodore-Amiga, while all the key players at Amiga enjoyed a significant windfall. For Miner and Morse, who had sunk so much of their own money into the Amiga project, this was particularly gratifying. The hard-working engineers received a lump of cash for their smaller shareholdings, along with a shiny new office in Los Gatos, fulcrum of the Silicon Valley. Suddenly, they had plenty of room, a big kitchen, and windows.

This was a smart move by Commodore as it gave the Amiga engineers good reason to stay on and finish the project. And there was still much to do, with a disk operating system being top of that list. This was supposed to have been in development for months, but the company contracted to do the work now wanted to renegotiate: it's one thing to agree a fee with a startup, quite another with a billion-dollar giant such as Commodore. In short, they wanted more cash.

At around this point, a Brit named Tim King enters the story. King had established himself as an expert on the Motorola 68000 processor and had already ported a Cambridge University-born operating system called TRIPOS (TRIvial Portable Operating System) to a 68000 computer. His company, Metacomco, saw an opportunity, and a particularly persistent colleague of King's managed to set up a meeting.

'I pulled up on the doorstep saying I've got an operating system that works on the 68000,' says King. 'I can probably move it over to the Amiga. And they were quite interested.' He picked up a hand-built Amiga and then flew it back to the UK in the seat next to him. 'I took it home and worked the hardest I've ever worked in my life. Literally, I would go to bed at 2am and get up at 6am working on this port.'

One of the biggest challenges was making the screen work. 'I spent days and days trying to get a picture on the screen and text on the screen,' he says. 'One night I phoned up California and was chatting to the guy who had written it. And I was saying I've done this, I've done this and I can't get the text on the screen. And he said, "What colour are you using?" I said, "Colour? I can only afford a black and white monitor. I've no idea what colour it is."' King laughs at the memory. 'He goes "oh", quietly to himself. "I may just have made it so that the default colour is invisible on a black-and-white screen."'

After around three weeks with the machine, King flew back to California with a floppy disk in his hand. 'I had to hang around in the morning and Bob Pariseau couldn't see me and said the programmers are very busy. "We can just about spare two people to come and see you and answer your questions," he said.'

With two engineers in attendance, King started to show what he had done. 'I put the disk into the machine and it boots up. And, of course, the Amiga had never booted up before, it had always just been programs inserted into memory. So I booted up, showed how to bring up windows, moved multiple windows around, showed how I could compile a program on it, showed how I could compile a program whilst editing a program at the same time, etc. And after about half an hour I'd run out of my little presentation and I turned around I found the entire company in a horseshoe behind me. And they all burst into applause.'

Not everyone was happy, most notably Carl Sassentrath, creator of the Amiga's kernel. Rather than adapting a disk operating system essentially written for mainframes, he wanted something faster, tuned for 16-bit processors and designed to work with floppy disks from the ground up. 'To which Bob Pariseau said, "look, it works and we haven't got anything else so this is it,"' recalls King. He, along with his wife Jessica who would also write much of the documentation, stayed for another three months to tune the operating system, and continued to work with Commodore for many years.

The world expected Commodore to announce a working Amiga at the January 1985 CES, but its engineers were battling problems on many fronts. The biggest one was yield of the three chipsets: Agnus, Denise, and Paula. 'By February 4, 1985, Denise yielded 50%, while both Agnus and Paula yielded 0%,' wrote Brian Bagnall [12]. It took them another four months of debugging to boost these to production-ready levels.

This turbulent period also saw Dave Morse leave Commodore-Amiga. Commodore head office had riled Morse by appointing Rick Geiger to shepherd development of the Amiga and report back on progress, and Morse finally left in May 1985.

The good news was that the company was, at last, in a position where it could ship products to developers. The bad news? The planned 128kB of RAM was nowhere near enough, which meant a redesign of the motherboard to accommodate 256kB of RAM and the option for adding another 256kB via an expansion slot.

One reason why the Amiga needed so much RAM was the ambitious operating system. Or rather, combination of operating systems. Quite aside from the disk OS created by Tim King, RJ Mical spent over seven months working on a graphical user interface called Intuition – at one stage, he was proud owner of a business card with the title 'Director of Intuition'.

Even then, it took a monumental effort from the wider Commodore team to push the first Amiga computers across the line. The company had committed to a glitzy launch at the Lincoln Center in New York on 23 July 1985, so a number of engineers were taught how to use C and given sections of code to work on.

Commodore had sent an early Amiga prototype to Electronic Arts, who rewarded it with the powerful bitmap graphics editor Deluxe Paint. Bravely, Commodore decided to use this as part of its on-stage demo to showcase the Amiga's powers. Even more bravely, it passed the mouse over to Andy Warhol as he used Deluxe Paint to give some Warhol-esque finish to a photo of Blondie's Deborah Harry. The Commodore engineers all knew that the fill command was buggy, and Warhol wasn't supposed to use it. He did anyway, but by some miracle it worked.

Commodore expected the watching press to write gushing coverage, but it had been burned by glitzy demos in the past: the first Apple Macintosh, in particular, had not immediately lived up to its on-stage hype. 'While initial reviews praised the technical capabilities of the Amiga, a shell-shocked PC industry has learned to resist the seductive glitter of advanced technology for its own sake,' wrote Fortune magazine [13]. Strange as it might seem now, many questioned whether multitasking, multi-channel audio, and an attractive interface were genuinely useful features for a personal computer.

If Commodore had been able to put enough Amigas onto shelves, however, it's likely the public would have quickly decided for themselves. Instead, financial problems and production delays meant the company badly struggled in the lead-up to Christmas 1985, and it only sold around 35,000 units. The Amiga's momentum was further hampered by high prices: over time the Amiga had lost its low-cost roots and moved into the top-end territory occupied by the Apple Mac and IBM Personal Computer. While it compared well to the Macintosh (a 256kB Amiga cost $1,295, albeit without a monitor, to $2,495 for the 128kB Mac), the Atari 520ST cost $599.

It didn't help that Commodore at this point was being led by a chief executive, Marshall Smith, whose steel industry origins and lack of interest in technology meant he didn't understand the computer market and could be easily swayed by people he really shouldn't listen to. The best example of this was his cancellation of the company's LCD portable computer project, being led – with little support from management – by Commodore engineer Jeff Porter. 'We had initial orders from Sears for 50,000 or 100,000 units,' Porter said [14].

Then Smith met the CEO of Radio Shack, who – according to Porter – told his Commodore rival, 'There's good news and bad news about the laptop market. The bad news is there is no market for this stuff. The good news is we own it.' Smith consequently canned the project.

Commodore's advertising for the Amiga was also poorly judged, attempting to out-Apple Apple with dystopian imagery that did nothing to sell the computer's unique features. And just to rub salt into its wounds, Jack Tramiel's technically inferior Atari 520ST was reportedly outselling it by a factor of ten (this turned out to be a big exaggeration; in fact, by 30 June 1986, Atari had sold 150,000 ST computers, which was roughly double the number of Amigas sold).

Many of the original Amiga team were upset by Commodore's handling of the Amiga release. One disgruntled engineer, who has never been named, wrote a message, 'We made Amiga, they f***ed it up' (the stars are ours). Without RJ Mical's intervention, this message would have appeared after Kickstart 1.2, the Amiga's

BIOS equivalent, had completed its work. Mical told the engineer to remove it, but he actually tweaked the code so that it would only appear for a 60th of a second.

This made the message invisible to the human eye, but it was caught on videotape and brought to the attention of Commodore senior management. Commodore withdrew all affected Amigas from sale, and in doing so – according to David Pleasance, then Commodore UK's marketing manager – delayed the launch of the Amiga in Britain by three months.

In the end, Brits had to wait until May 1986 for the official release of the Amiga in the UK – and were immediately taken aback by its price of £1,475 excluding VAT. With tax, it came to just shy of £1,700. This did not compare well to the newly announced Atari 1040ST, which included 1MB of RAM and a colour monitor for £899. Still, Commodore managed to sell some Amigas to British schools and colleges. 'We also did well in the audio-visual market,' said Pleasance [15]. 'Where we didn't do well is in general business, word processing, and stuff like that, because it was a lot of money for something that other machines for a lot less money could do.'

Kelly Sumner, who worked for sales in Commodore UK at the time but who later became general manager, described the market as 'tough, very tough' [16]. He estimated that the UK sold around 14,000 of the first Amigas in twelve months. It didn't help that Commodore still didn't seem to understand what the Amiga was. For instance, when a review sample finally arrived with Personal Computer World it was supplied with the Sidecar: a bulky expansion unit that cost around £750 and turned the Amiga into an IBM PC-compatible machine. So, the price of an IBM clone.

As British magazine Your Computer noted in its June 1986 article, 'Amiga – the future is here' [17], the only way the Amiga could justify its high price was due to the staggering amount of power it contained. However, it needed software to take advantage, and aside from EA's Deluxe Paint this was still largely absent. Commodore's own Musicraft looked promising but was still being developed: 'When I first heard the stereo version of *Axel F*, using sampled sounds, even an Amiga aficionado like me was truly amazed,' wrote Your Computer's Francis Jago. He also praised Animator from IEF Aegis for allowing even 'the non-programming user' to create compelling animations. But three good programs isn't enough to sell a computer.

Then there was the lack of games to consider. With so few sales of Amigas by the end of 1986, it would take a bold developer to switch its attention to this new, unproven platform. Commodore needed to produce a cut-price version of the Amiga if it was going to succeed in the mass market, the company's traditional strength.

However, this would mean compromises to the Amiga's aesthetics and philosophical approach. As such, it's unsurprising that it took a team from outside of the original Amiga group to come up with a workable plan for a low-end Amiga; indeed, Miner's team was working on an even more expensive version of the Amiga codenamed Ranger.

What eventually became the Amiga 500 (or A500) started life as a project called B52 on 9 May 1986. 'We nicknamed the B52 after the bomber, not the rock group,' said Jeff Porter [18]. 'We either have to bomb the competition or we might as well bomb ourselves because we're going down in flames if we don't hit a home run here.' The team's budget for materials was set at $200 with an original target launch date of late October; both inevitably slipped, but it was enough to fire up the project and convince Commodore's new boss, Thomas Rattigan, to sign it off.

The first step was to redesign the case to make it more like a C64. After two aborted attempts – first to use Rob Gemmell, designer of the Apple IIc and IIe, then some radical designs using a student at the local Philadelphia College of Art – the eventual design came from Commodore Japan's Yukiya Itoh. While the A500 controversially regressed to a single unit with an integral keyboard, it crucially cost less money. This was also why the A500 has an external power supply to the A1000's internal unit.

Neither of these decisions sat well with the original Amiga team, who were still based in Los Gatos and unhappy with the direction the B52 project was taking. At one point, a fax war broke out, with Miner's team sending one explaining what Porter's team was doing wrong and Porter then sending back a point-by-point riposte. He eventually decided to visit them in California to go through everything in person, and while it would be lying to say that tensions between the two teams disappeared, Porter's experience and fact-based approach eventually won them over.

As momentum on the B52 project built, so did team spirit. The Amiga's rebirth had all the hallmarks of early Commodore efforts, with a group of passionate engineers working stupid hours to get the product out of the door. All the while, Porter was looking for ways to save money. He spotted a 16-bit wide ROM chip in Nintendo NES cartridges and used that to store Kickstart; he slashed the cost of the PCB (printed circuit board) by switching from a four-layer to a more primitive two-layer design; he dropped features such as support for genlock, which was necessary for some of the original Amiga's high-end video effects. Output to TVs disappeared too.

But it wasn't all about cost-cutting. 'Fat Agnus' addressed 1MB of video memory compared to 512kB of the original Agnus, while the A500 would also launch with a much more stable operating system thanks to the continuing work of Miner's Los Gatos software engineers.

With the launch fast approaching, Gerard Bucas – the Commodore executive in charge of making the B52 project happen – put his job on the line by signing off on final tooling. This was expensive (think five zeroes) and Henri Rubin, who was chief operating officer and had been instructed by Irving Gould to take control of spending, wouldn't sign off on it. 'The best thing we can say about [Rubin] is that he was tremendously good at never putting his signature on anything, never making a decision and just generally delaying things,' was Bucas's biting verdict [19].

Porter had hoped to launch the Amiga 500 in a blaze of glory at the January 1987 CES, but delays to the production of the Fat Agnus chip meant it was only shown to a select group of journalists and dealers in a private booth. But Commodore did confirm the name, the price ($650) and that it would include 512kB of RAM to the original Amiga's 256kB.

The good news for Europeans is that they didn't have to wait months after the American release. In fact, this was almost simultaneous, with Commodore International using the European trade show CeBIT in March 1987 to announce the A500 and the high-end A2000 (see 'The A2000 story').

By the time Commodore produced its 1987 annual report for shareholders, the company declared: 'Market response to both new computers has been strong. These products will provide a solid platform for sales growth in the coming year.' And, if nothing else, they solved the split personality problem that afflicted the original Amiga: the A500 was for a home user and priced as such; the A2000 was for businesses, and with prices to match there too.

The same report boasted that more than 700 titles were now available for the Amiga range, including 'many popular business programs' such as WordPerfect. This word processor was an undoubted coup, especially bearing in mind how basic Commodore's own offering was, but the computer press still felt the Amiga's software range was lacking.

In February 1988, for example, The Guardian published a story lamenting the lack of software for the computer. 'Software houses will only write for machines that sell,' wrote Christine Erskine [20], an Amiga owner, 'and despite the fact that the Amiga sold reasonably well last autumn and Christmas (up to 30,000 units on a best estimate), the development costs and small user base mean that for publishers to invest heavily in the Amiga is a risky business.'

Erskine made two other key points: the Amiga 500's arch-rival the Atari 520ST had appeared on the shelves a year earlier and was cheaper (at that point, £399 to the Amiga 500's £599). It therefore made sense for software companies to write for the Atari first and then port their games over, which meant that graphics and sound weren't as good as they should be as they were writing for the lowest common denominator. Her other point: software makers no longer operated in an 8-bit world. To create something that used the full abilities of even the Atari, it took time, resources, and therefore money.

Some games did showcase the Amiga's power, though, with *Defender of the Crown* (released in 1986) being the most obvious example. RJ Mical wrote the game's engine and – although he wasn't credited at his request – his intimate knowledge of the Amiga's chipset was obvious. It took full advantage to create the closest thing the world had seen to photorealism in a game, up until that point.

Not that Mical wants to take all the acclaim. 'Jim Sachs, what a god he is,' said Mical of the game's lead artist [21]. 'Jim Sachs is amazing. These days everyone sees graphics like that because there are a lot of really good computer graphics artists now, but back then, 20 years ago, it was astonishing to have someone that good.'

Defender of the Crown set the benchmark for Amiga games, and as sales of the A500 grew so did the number of high-quality games to play. Perhaps most famously, *Lemmings* made its debut on the Amiga after Mike Dailly of the Dundee-based DMA Design took up his own challenge of creating 8×8-pixel men using Deluxe Paint.

From this point on, the Amiga followed a similar life story to a typical *Lemmings* level. Along with the many squashed disappointments – the misguided attempt to bake in IBM PC compatibility, the lack of success at appealing to the general business market – there were many level-up triumphs. This was particularly true in traditional Commodore strongholds of Germany and the UK, with estimates suggesting that both countries sold 1.5 million machines out of the Amiga's 5 million total.

Pleasance puts much of the Amiga's success down to the bundles he and his team put together. 'I went to Ocean and met up with David Ward and Jon Woods,' he told the British Computer Society in 2019 [22]. 'I said, "I'm going to put a proposition to you. You'll either have the balls to go through with it or you'll have me taken away." ' That proposition was to bundle a copy of *Batman the Movie* with new Amigas and, while such bundles are commonplace now, there was initially resistance from computer retailers. After all, they would only benefit from the sale of the computer, not the software. Ocean Software was also concerned that individual sales of games may be affected.

'And yes, we did affect Ocean's sales: they ended up selling five times more copies than their biggest estimate of sales," said Pleasance, adding that he ended up ordering 186,000 copies of the game. 'That's how many Amiga 500 *Batman* packs we sold in twelve weeks.'

Not that people only played games with their Amigas, as we discovered in our survey of 1980s computer users. Two responses in particular show how influential this computer was on Britons in the late 1980s and early 1990s. 'It wasn't dry like PCs were at the time,' Dominic Reid writes. 'It was fun. I am a professional software dev now because of Easy Amos on the Amiga.' David Rintoul worked in software design and support for over 20 years thanks to his Amiga experiences: 'Superb operating system to program. My first experience of multitasking and I think it was object oriented too.'

With solid sales figures, and such praise from its advocates, it's wrong to describe the Amiga as a failure. However, there's always a feeling that it could have done better. So why didn't it?

At the very start of the Amiga story, we started with a series of maybes. While those chips fell in the Amiga's favour, a series of poor management decisions – and a simple lack of cash to invest – meant the Amiga never had a proper chance to thrive under Commodore. Could it have outflanked the IBM-compatible PC? That seems unlikely, despite its technical superiority. Could it have usurped Apple? Quite possibly. Or it might simply have given us a third major platform with its own dedicated set of users.

The truth is we will never know, but the Amiga does live on. Not just in memories but in events, online resources – and in the thousands of A500 and A2000 computers that are still working, beeping, and bouncing big red-and-white balls around the screen.

The legend of Mitchy

On the release of the original Amiga, all 53 members of the team involved with its release signed the mouldings for the inside of the case (an echo of the Macintosh). One of the signatories was Jay Miner's black cockapoo, Mitchy.

'Miner had a brass nameplate on his door that read, "J.G. Miner", and just below it was a smaller nameplate, "Mitchy",' wrote Brian Bagnall in his comprehensive tome, *Commodore: The Amiga Years* [23]. 'The canine even had her own tiny photo-ID badge clipped to her collar as she happily trotted through the halls.'

'There were four beings always in our discussion,' said Ron Nicholson [24]. 'It was Jay, Joe, myself, and Mitchy, sometimes napping but sometimes carefully observing the proceedings. Occasionally we had to figure out which way to go, left or right. We'd toss it up and say Mitchy, what do you think? One bark for yes, two barks for no.'

While RJ Mical was a big fan of Mitchy, even keeping dog treats in his drawer, this wasn't a feeling shared by everyone. 'I never met Jay Miner but I have a lot of respect for him,' said Bil Herd [25]. 'The only thing I ever heard bad about him was that his dog stunk.'

The A2000 and the Video Toaster

The original Amiga A1000 was effectively replaced by two computers: the A500 for home users and the A2000 for businesses.

The A2000's key aim was to run two operating systems natively: Amiga's OS, including its graphics user interface Intuition, using its Motorola 68000 processor, and MS-DOS thanks to an optional PC compatibility card called the bridgeboard;

according to Tim King, 'this was initially called "Janus" after the two-headed god'. Add the bridgeboard and you could also run business software such as Lotus 1-2-3.

Expansion was another key feature. The basic unit, which cost $1,495, included an integrated 3.5-inch floppy drive, but buyers could add a second 3.5-inch floppy drive and a 5.25-inch floppy drive in spare bays at the front. Then there were all the internal expansion ports for adding endless types of cards: network, graphics, CPU, memory, serial port, SCSI adapters and, of course, the bridgeboard (which included an Intel 8088 processor).

'The technical complexities are enough to make your eyes water,' wrote Jack Schofield in The Guardian on the A2000's launch [26]. 'Imagine doing a simple task like collecting a keypress from the keyboard. Is this intended for the Amiga half? If not, how can it be channelled to the PC half, which uses a different type of keyboard altogether?'

Schofield goes on to explain the 'convoluted process' involved, but his key point came at the end of the article: that you could buy both a 68000-based computer and an IBM-compatible PC for less than the price of the Amiga 2000 'and still have change to spend on software'.

While the A2000 wasn't a mainstream hit, it helped to reinforce the Amiga's credentials for creating video. Key to this was an advanced port that designer Dave Haynie added to the design, after he was instructed by Commodore chairman Irving Gould to take over the design from Commodore Germany.

'The original German A2000 has a 'genlock slot', which was basically just an internal card edge version of the external 23-pin Amiga video port,' commented Haynie on an Ars Technica article about the Amiga (which he was correcting) [27]. 'I added an extra connector to deliver the full 12-video video output. George Robbins... was helping me out, and he suggested running the A2000's parallel port to that slot as well, so we'd at least have some way to control things on that slot... those two decisions enabled the Video Toaster and that, for pro video on the Amiga, was all that mattered.'

In some people's view, the Video Toaster was just as important to the Amiga as VisiCalc was to the Apple II: that much-abused term of a killer app. 'It really did enable the Amiga to make it in the digital video world, much like the early Mac conquered by inventing desktop publishing,' says Tim King. As a result, the

Amiga 2000 and its successors were an important part of professional video creation for years to come.

Sources

Interviews with Neil Harris, Tim King, and Leonard Tramiel.

1. Scott Stilphen, DP Interviews… Larry Kaplan, Digital Press, 2006
   digitpress.com/library/interviews/interview_larry_kaplan.html

2. A. Stuart Williams, *The AUI Interview: Jay Miner – The father of the Amiga*, Amiga User International, 21 January 2018
   web.archive.org/web/20201031211301/amigauserinternational.com/2018/01/21/the-aui-interview-jay-miner-the-father-of-the-amiga

3. John Chambless, *Making Sense of High Tech*, Chester County Press, 6 June 2018
   chestercounty.com/2018/06/06/174738/making-sense-of-high-tech

4. Brian Bagnall, *Commodore: The Amiga Years*, Variant Press, Kindle Edition, location 2411

5. John Chambless, *Making Sense of High Tech*, Chester County Press, 6 June 2018
   chestercounty.com/2018/06/06/174738/making-sense-of-high-tech

6. John J Anderson, *Amiga Lorraine: Finally, the 'Next Generation Atari'?*, Creative Computing, April 1984, page 150

7. Brian Bagnall, *Commodore: The Amiga Years*, Variant Press, Kindle Edition, location 2887

8. As above, same location

9. As above, location 3026

10. As above, location 3090

11. C.W.Miranker, *Atari accuses Amiga of breach of chips contract*, D7, San Francisco Examiner, 21 August 1984

12. As above, location 4659

13. Jeremy Reimer, *A history of the Amiga, part 4: Enter Commodore*, Ars Technica, 22 November 2007
    arstechnica.com/gadgets/2007/10/amiga-history-4-commodore-years/4

14. Brian Bagnall, *Commodore: The Amiga Years*, Variant Press, Kindle Edition, location 4875

15. As above, location 7020

16. As above, location 7032

17. Francis Jago, *Amiga – the future is here*, Your Computer, June 1986, page 56
    archive.org/details/your-computer-magazine-1986-06/page/n55/mode/2up

18. Brian Bagnall, *Commodore: The Amiga Years*, Variant Press, Kindle Edition, location 7233

19. As above, location 8478

20. Christine Erskine, *Those unfulfilled aspirations*, The Guardian, 11 February 1988, page 27

21. Jeremy Reimer, *A history of the Amiga, part 7: Game on!*, Ars Technica, 5 December 2008
    arstechnica.com/gadgets/2008/05/amiga-history-part-7

22. Martin Cooper, *The rise and fall of Commodore*, BCS.org, 2 July 2019
    bcs.org/content-hub/the-rise-and-fall-of-commodore

23. Brian Bagnall, *Commodore: The Amiga Years*, Variant Press, Kindle Edition, location 879

24. Brian Bagnall, *Commodore: The Amiga Years*, Variant Press, Kindle Edition, location 1130

25. Brian Bagnall, *Commodore: The Amiga Years*, Variant Press, Kindle Edition, location 3843

26. Jack Schofield, *Too clever by half*, The Guardian, 5 March 1987, page 23

27. David Haynie, comment to *A history of the Amiga, part 6: stopping the bleeding*, Ars Technica, 5 January 2019
arstechnica.com/gadgets/2008/02/amiga-history-part-6/?comments=1

Amstrad
PCW 8256

The typewriter killer

You might have thought that Alan Sugar would, by February 1985, have learned to keep a notebook to hand so he could scribble down his product ideas. Instead, he used the back of a Cathay Pacific serviette to sketch out the inspiration behind the PCW series of all-in-one word processors. It was a moment of inspiration that would earn Amstrad tens of millions of pounds.

In retrospect, the idea itself seems so simple. IBM was then selling its DisplayWrite word processors for around £5,000. What if Amstrad could create an all-in-one word processor that combined a keyboard, floppy drive, screen, and printer in one tower system? And what if it could do so for a tenth of the price?

While Amstrad ended up making some compromises, the final system proved to be exceptionally close to Sugar's original, serviette-based concept.

Before Amstrad could develop the idea, though, it had one big problem. Who was going to supply the printer mechanism? Even in the mid-1980s, this was specialist technology that only a handful of American and Japanese companies had mastered. Would any of them be willing to share their secrets so that Amstrad could create a fully integrated word processor?

To find out, Sugar organised a meeting with Seikosha at its head office in Tokyo. (Seikosha is better known by its export brand, Epson, outside of Japan.) He explained the concept behind the all-in-one word processor to an audience of two men: one senior, one young and quiet. He emphasised that he had no interest in building a printer himself; he simply wanted a printer mechanism. Oh, and the software codes to make the printer work. 'As is typical of Japanese meetings,' wrote Sugar in his autobiography, *Alan Sugar: What You See Is What You Get* [1], 'they said they would come back to me, but when I left I didn't feel very confident. Without a printer mechanism, we'd never get a word processor off the ground, so I virtually ditched the idea.'

But – and this is a common theme during the creation of the PCW 8256 – good fortune was on Amstrad's side. Shortly after he returned to London, Sugar received a call that two gentlemen representing Seikosha were keen to meet up with him. It turned out that the quiet, young man who he had met was both the son of Epson's boss, keen to impress, and enthusiastic about the Amstrad idea. 'I wrote them an order there and then for 100,000 pieces,' wrote Sugar. 'What a lunatic I was in those days!'

With the printer mechanism sealed, it was time to set his computer-manufacturing army into motion. 'It started with a two-page fax from Alan Sugar,' remembers Mark-Eric Jones, better known as Mej, who was the circuit board designer behind the original Amstrad CPC 464. 'I remember very clearly one sentence in it, where it basically said, "Clear your mind of all standards. We want this to be very optimised, don't try and just do things in a standard way. We want to create the best possible product of its type." '

The other convenience of forgetting all standards was that it cut costs. At that time, printers needed their own processor; Amstrad would effectively turn the printer into a semi-dumb box, with the Amstrad PCW's processor sending the print instructions directly through its proprietary cable. This meant the printer unit could be simple and cheap. 'We ended up, I think, with a printer with just two or three chips in it; a very cost-efficient design,' says Mej.

Those chips were crucial, however, with a microcontroller and integrated circuit – both designed in-house by Mej and his team – that could interpret the command from the main computer and send the right information to the printer's mechanics. 'Technically, the mechanics and timing of the printer was one of the most interesting challenges,' says Mej. 'Linking the printer and the low-level software so that you could send the command down the wire to the printer, and it would then go and fire all the needles with all the right timing, then ramp up the motors to move the carriage, all those things were being controlled.'

As with the CPC 464, Amstrad used its injection moulding expertise to create a striking design for the monitor. Although, if it had kept to the original design, it would have been more striking still: the first concepts included a portrait screen, not landscape. 'Mass-produced electronics don't really allow you to do portrait screens,' explains Roland Perry who, as with the CPC 464, was project manager for the PCW. 'It's all TV technology and TV technology is landscape not portrait.'

Crucially, however, this screen would be much higher resolution than its competitors. 'You couldn't see the whole A4 page on the screen at once, you had to scroll down,' says Perry, 'but at least you could see the whole width of it because it was 90 characters wide.' This was an immediate selling point for the PCW when it went on sale. 'Typically with word processing packages on general-purpose computers at the time, 64-character screens would drive you crazy because you're scrolling left and right all the time.'

Users also benefited from a higher-resolution screen and more detail as a result. Even if you were fortunate enough to be using an 80 by 24 screen until now, you'd relish the extra detail afforded by the PCW's 90 by 32 display. In fact, you'd have 50% more characters on the screen. Nor did this higher-resolution screen add to the cost, with Amstrad choosing what Perry describes as a 'pretty bog standard' cathode ray tube.

The team behind the PCW 8256 resisted any temptation to make huge changes to the core components that had served them so well for the CPC 464. Everyone involved was familiar with the Zilog Z80 processor, which was more than powerful enough for word processing tasks. The major upgrade was to memory, with 256kB as standard rather than 64kB; this wasn't a frivolous upgrade driven by a marketing team keen on big numbers, but a necessity to store the word processing program, to drive the screen, and to store the document being worked upon.

The team designed the circuit board from scratch, with Perry describing the early prototype as 'a stack of two boards'. Each board was A4-sized and stuffed with logic chips. 'Mej's job was to take that logic, that amalgamation of TTL [transistor-transistor-logic] gates and all the rest of it, and turn that into a single custom chip,' says Perry. 'But by then we were fairly familiar with this concept that you could take about a square foot of discrete logic chips and turn it into one single chip.'

With prototypes to work on, Locomotive could start on the key job of writing the word processor software. And here's where a second dollop of good fortune lands: the Locomotive team had years of experience of doing just that, because they had built standalone word processors for Data Recall – a company owned by Mej's father.

This experience meant they didn't need to teach themselves the ins and outs of creating a word processor, but instead concentrate on making LocoScript the best word-processing package around. And they had the advantage of knowing all the details of the printer mechanism that would be printing the final results. 'I think one of Locomotive's greatest contributions was to go, "You may say this is an 8-pin printer mechanism, but if we just advance it by one-third of the normal notch each time, we can turn this into a 24-pin output."'

Admittedly, Locomotive wasn't the first company to have this idea, but users wouldn't care. It allowed the Amstrad PCW to offer two print speeds: one for quick and dirty drafts, where you just want to read the words yourself, and the other for higher-quality output when you might be writing a formal letter or creating a presentation.

Locomotive's cleverness extended to formatting too. While existing software word processors could add professionalism to documents by justifying right, the text would become jagged when you started playing around with formatting such as bold, italics, or extra-wide, extra-big text. With a PCW, you could add whatever formats you wanted and it would still be perfectly justified.

The final ingredient? Floppy drives. These weren't a new development for Amstrad, with the CPC 664 – introduced in May 1985 – already including the 3-inch drives that would become Amstrad's trademark. As Perry describes in 'The 3-inch floppy story that won't go away' on page 267, Amstrad even had a prototype 3-inch external drive that it used when the CPC 464 was first demonstrated.

Integrating the floppy drive into a monitor housing was a first for Amstrad, but Perry insists that this was a relatively easy task. 'Bob Watkins's people would have produced that very quickly,' he says, referring to Amstrad's technical director and, in effect, Lord Sugar's right-hand man. 'Getting injection moulding done was one of Amstrad's core skills.'

The original plan was for the PCW to ship without a full-blown operating system, instead including a version of BASIC to accompany the word processor. 'Very late in the day, we had pressure from some of the distributors, particularly Schneider in Germany, who said, look, you might think this is a word processor, but we think this is a CP/M engine.' In short, they felt they could sell far more of the machine if it was a general-purpose computer based on the then-popular operating system (see the story of the IBM Personal Computer).

At the same time, software companies were protesting that Amstrad was going to sell a major computer and not allow them to write software for it. 'Right at the end, I convinced everybody that rather than have a third-party company doing, say, a not very good CP/M Plus port onto it, we should do our own port of CP/M Plus and therefore head that all off,' says Perry. 'If people want to use it effectively as a 6128 with a big screen on it and a printer then they could.'

The PCW made its debut, complete with CP/M Plus, at a typically over-the-top press event in September 1985. To directly quote David Thomas in *Alan Sugar: The Amstrad Story*[2]: 'Three actresses represented three different types of secretary: a frightfully snooty one would not dream of using anything less than a £10,000 word processor; a tarty secretary swore that a typewriter was good enough for her; while the cool, efficient secretary, of course, preferred the Amstrad word processor.'

'I thought the launch was schmaltzy,' says Rupert Goodwins, who was then working for Sinclair but, having written a few articles for British computer magazines, had sneaked in using press credentials. 'But I thought the product was astonishingly good. Just absolutely one of the most impressive bits of 1980s computer technology the UK produced.' Goodwins would report back to Sir Clive that he couldn't believe what Amstrad had produced for the money. 'The fact it was the monitor and the printer and the CPU and the disk drive and all that for £399 – it was just such an astonishingly clever design.'

Early press coverage from Personal Computer World magazine [3] was equally positive, declaring it 'spectacular value for money' and stating that 'it's difficult to find any competition' when viewing the PCW 8256 as a word processor due to its price, while pointing out that, even if you consider it as a CP/M machine, there was 'no contest'. Its verdict was decisive: 'How can you criticise a machine that gives you 256k of RAM, a disk drive, a monitor, a printer, a very good word processor, BASIC and Logo for £399 plus VAT, even if it does use your name?' (Personal Computer World was usually fondly referred to as PCW by its readers and writers.)

To say the PCW 8256 sold like hot cakes during the lead-up to Christmas 1985 reflects rather too well on the attractiveness of hot cakes. The challenge was to find a PCW available to buy, with Perry describing how Dixons broke one of its own mantras: 'Their golden rule was never let the customer leave the shop without having bought something. Even if the thing they came in to buy wasn't in stock, you're supposed to sell them a reasonable substitute. They couldn't do that with the PCW because it was unique.'

When stocks arrived, Perry describes a 'bush telegraph that went around as people found out. They'd say, oh, the Dixons in Manchester just had a delivery of PCWs and everyone would literally drive around there. By lunchtime, they'd be sold out.'

The PCW series would go on to sell 8 million units worldwide, making it one of the most popular British-made computers in history and earning Alan Sugar a small fortune in the process. It would prove to be a hit for homes, small offices, and university students thanks to its low price, with Sarah Kidner – a journalist and editor – using it to write her dissertation and fondly describing it as a 'great word processor'.

The PCW 8256 even travelled the oceans. Richard Gough, Chief Technology Officer of the Royal Navy, described how he 'secured the system with an anchored custom-made stand so it would remain in place in rough seas' [4]. Not content with its word-processing skills, he also created an HR system to keep track of training, starters, and leavers.

With such a proven hit in the UK, it seems odd that the all-in-one word processor/home computer didn't have a greater worldwide effect. But Roland Perry does have a theory: 'The PCW struggled outside the UK because it was 8-bit, a dirty word by that time. People had become very rapidly conditioned to only want to buy 16-bit, and preferably something which hinted at IBM compatibility.' In particular, the PCW struggled for sales in the USA, which had proven to be fertile ground for the CPC series.

This didn't stop Amstrad from releasing a succession of follow-up machines, with the PCW series even making the jump to 'mobile computer' with the NC 100, NC 150, and NC 200 models (see page 269).

For Amstrad, though, the big hope was its IBM-compatible PC. As described in the story of the IBM Personal Computer, anyone could build a 100% compatible computer by reverse-engineering the BIOS, and once again Mej came to the rescue by doing precisely that. Amstrad knew it could produce a much lower-priced system by simplifying the circuit board, and it also struck a deal with Digital Research to run its GEM graphical interface.

Once you mix in Amstrad's usual aggressive pricing strategy – its range started at £399, but even its most expensive machine (with a 20MB hard disk) only cost £949, compared to £1,429 for an IBM PC – Alan Sugar felt confident enough to boast that he would sell 70,000 units a month.

But then Amstrad hit two big problems. The first was supply of the hard disks, the second a rumour that its PCs were unreliable due to overheating. This rumour gained traction because Amstrad had built the power supply into the monitor, allowing it to remove a cooling fan from the base of the computer. ICI supposedly had rejected the Amstrad PC for this very reason and, even though it publicly refuted this claim, Sugar eventually bowed to pressure.

Although he did it in his inimitable way: '... if it's the difference between people buying the machine or not, I'll stick a bloody fan in it,' he told a journalist who enquired about the overheating rumours. 'And if they say they want bright pink spots on it, I'll do that too.' [5]

The killer blow to Amstrad's computer fortunes came when it produced the 2000 series, with the first batch needing to be recalled due to hard disk failures. This was proved to not be Amstrad's fault, rather the hard disk manufacturer, but the mud stuck and Amstrad PCs gained an unjustified reputation for being unreliable.

But it's wrong to end on a downbeat note. Amstrad's designers came up with some of the most interesting and innovative computers of the 1980s, and it is also the only British computer manufacturer that broke through in the USA. Not bad for an East Ender who started his business with a cheap van and nothing in his bank account.

What's in a name?

Every new computer has a codename during its development process and William Poel, who joined Amstrad (along with Roland Perry) after the success of the CPC 464, decided to dub the all-in-one word processor 'Project Joyce'.

It was a fitting name: Joyce Caley was Alan Sugar's personal assistant at the time, and she would sit guarding his office on the top floor of the company's Brentwood headquarters. You weren't getting in to meet the great man without her say-so. Crucially, so far as the PCW 8256 was concerned, Joyce was skilled at shorthand and typing, so would handle all correspondence for the Amstrad boss.

'If you ever watch *The Apprentice*, the ever-shrinking person who sits in the lobby, who answers the phone call where he says, "You can come in now," is Joyce – figuratively,' says Perry. But Alan Sugar didn't intend for Joyce to be replaced by a computer. 'His vision was that anybody could do the shorthand, typing part of the job themselves. And if they didn't want to do it themselves, they could give their version of Joyce one of the word processors and they could then use that instead of a typewriter and therefore be more productive.'

The PCW 8256 also came perilously close to being called the WPC 8256, short for word processing computer. Right to the point where a front panel was etched and the sticker applied. 'I said, you've got to be joking,' recalls Perry. 'Woman police constable? 'Ello, 'ello, 'ello, what's going on here then? I said that's just ridiculous. And they [Bob Watkins and Alan Sugar] looked at me and they said, "Oh, I suppose you're right." So they changed it to PCW.'

The 3-inch floppy story that won't go away

If you want to irk Roland Perry, repeat the rumour that Amstrad opted for the 3-inch floppy drives because they were available at fire sale prices. 'That whole story makes no sense at all,' he says, his frustration palpable that the rumour still exists. 'Amstrad does not do production in quantity, let alone by the million, with bankrupt stock. Say there's 50,000 surplus Hitachi drives in the warehouse. That's going to be the first week's production and then what do you do?'

It seems likely that what Perry describes as 'the meme that won't die' stems from when Amstrad created a prototype DDI-1 drive, which did use soon-to-be-discontinued Hitachi drives. And it gathered weight because 3.5-inch floppy drives went on to become the dominant choice for portable storage.

However, the ascendency of the 3.5-inch drive happened after the CPC 464 and PCW 8256 went into production. 'One of the things that over the years has always rubbed me up the wrong way is this idea that there was something quirky about the way we chose the 3-inch disk drive for the 464 and the PCW, when "everybody else" was using the three-and-a-half inch desk,' said Perry. 'Well actually, nobody else chose it for several years afterwards. That's all just rewriting history.'

In reality, Perry had settled on the 3-inch floppy drives during the development of the CPC 464, even using that prototype DDI-1 drive in the press launch so that it could quickly load fully functional games like *Roland in the Caves* and *Roland on the Ropes*. 'This means we picked the 3-inch disk drive by Easter 1984; it wasn't picked a year later, when we launched the 664. We'd already made that decision. And it was absolutely the only candidate – floppy disk drive, Microdrive, or anything similar – which had any legs at all, when I looked at half a dozen alternatives.'

What came next

The PCW 8256 was phenomenally successful in the UK, and Amstrad was quick to follow it up with upgrades and successors. Not all of which matched the original...

PCW 8512

Release 1986   **Price** £499 plus VAT

Only a few months after the release of the 8256, the 8512 appeared with twice the memory and two floppy drives.

PCW 9512

Release 1987   **Price** £499 plus VAT

The 9000 series offered two key upgrades over the original PCW: a white-on-black screen rather than green-on-black, and a more sophisticated daisy-wheel printer to replace the dot-matrix technology of the original. Amstrad also shipped the updated LocoScript word processor, integrating a spellchecker and mail merge.

PCW 9256 and 9512+

Release 1991   **Price** £349 and £449 plus VAT

Amstrad finally moved away from 3-inch floppy drives with these two new models, choosing instead a 3.5-inch drive with a 720kB capacity. With the 9512+, you could also choose a more expensive Canon inkjet printer rather than the daisy-wheel, Amstrad-badged unit introduced with the 9512.

PcW16

Release 1994   **Price** £299

With its own graphical user interface – dubbed Rosanne – and a collection of utilities (word processor, spreadsheet, address book, diary/alarm, calculator), the PcW is an oddity among the PCW series. Add a garish keyboard that only its mother could love and success was always a long shot.

NC 100, NC 150, and NC 200

Release 1992 to 1993   **Price** £199 to £329

While the NC series of portable computers wasn't truly part of the PCW series, the idea was similar: create an all-in-one device with a built-in word processor. They still used the Z80 processor too. With an 80-column by 8-row mono LCD screen, the NC 100 could be powered for up to 20 hours by four AA batteries and included a serial port, parallel port, and PC card socket. The NC 200 added a flip-up screen and backlit 80×16 display, along with a 3.5-inch floppy drive, but needed five C cell batteries.

Sources

Interviews with Rupert Goodwins, Mark-Eric Jones, Roland Perry, Dick Pountain and Jack Schofield.

1. *Alan Sugar: What You See Is What You Get*, Pan Books 2011, page 253
2. David Thomas, *Alan Sugar: The Amstrad Story*, Century 1990, page 175
3. Peter Bright, Personal Computer World, October 1985, page 132
4. Richard Gough, *Tales of Early IT: Amstrad PCW 8256*, LinkedIn Pulse, 29 August 2019
    linkedin.com/pulse/tales-early-amstrad-pcw-8256-richard-gough [requires LinkedIn registration]
5. David Thomas, *Alan Sugar: The Amstrad Story*, Century 1990, page 230

Acorn
Archimedes

The beginning of the
ARM revolution

Project A. Its goal was easy to describe yet, even now, seems impossibly bold: to design a microprocessor*. How could a little computer company in Cambridge be so arrogant as to think that it could design its own processor? That was for the Americans, the Israelis, the Japanese, not little old Britain. And yet that chip was to become the ARM, a microarchitecture and instruction set that is now powering nearly every smartphone in existence.

It all started with Acorn director Andy Hopper, who led a dual life as both an entrepreneur and an academic at the Cambridge University Computer Laboratory. In June 1983, the University of California, Berkeley, had published its work on RISC I, 'a Reduced Instruction Set Computer'. Hopper had read the papers and immediately sensed their importance: 'I showed them to Hermann [Hauser, co-founder of Acorn], I showed them to Steve Furber. And I said, this is important. Could we do this?'

The timing couldn't have been better. By now, the BBC Micro was a booming success and had turned Acorn from a £1 million turnover company into a £30 million powerhouse. Its management team had equally big ambitions for its next computer, but key engineers Furber and Sophie Wilson had grown frustrated in their search for a processor to put at its core.

The duo had tested all the latest 16-bit chips to see exactly how fast they were, but kept on hitting the same problem. Whether it was the Motorola 68000, the Western Design Center 65C816, or the Intel 80286, 'they all performed the same,' says Wilson. 'We sort of coined a law that all these processors, no matter what claims have been made, all performed the same because they all had the same memory bandwidth – in particular, instruction memory bandwidth.

'We were quite upset, in a way, by the advanced processors, because we knew how to build a four-megabyte bandwidth memory system and we could have made that 32 bits wide and there was nothing out there that could use it,' adds Wilson. 'And we just theorised that such a machine would be really very quick.'

By the time Hopper discovered the RISC research, Acorn had effectively done a worldwide tour to find a processor for its next-generation computer. In particular,

* Well, unless you want to describe it to someone who knows about such things. You don't simply design a microprocessor. 'You also design the associated chips for I/O, memory, and video,' says Sophie Wilson. 'And write the software tool chain – assembler, linker, compiler – and end application software.'

they had visited National Semiconductor in Israel and Intel in California, and come away impressed by their facilities but disappointed by some of their decisions.

'We had a meeting with Intel saying, look, the 80286 is an OK processor, you just screwed up the pin-outs because you put both the address and the data bus on the same pin,' says Hauser. 'Nobody can make a good computer out of that. But if you sell us the die, we can do our own pin-out and maybe we can make something out of this. They told us to get lost and we said well you get lost, we'll do our own. It's the only reason why the ARM exists. If they had given us the 286 die, we would have used that rather than do our own ARM chip.'

At around the same time that Furber and Wilson were ruminating on what a RISC-based computer architecture might look like, Acorn decided that, for an unrelated project, it would take advantage of the 65C816 processor from Western Design Center. Time to send Furber and Wilson back to America, but this time to WDC's base in Phoenix, Arizona.

'We'd visited other companies,' says Wilson. 'In particular, we visited National Semiconductor several times about the 32016, which they claimed would be their next great scientific processor. And that was standard Silicon Valley fare of a big building with lots of engineers in it. We turned up at the Western Design Center, and it was two bungalows on a road in Phoenix, staffed by a couple of senior engineers and a bunch of grad students. Steve and I left with the overwhelming feeling that, hell, if they could build a processor then we could too.'

What made Furber and Wilson so convinced that the RISC approach was the right approach? 'I always say that RISC stands for Reduced Instruction Set Complexity,' says Wilson. 'You can have lots of instructions – ARM1 has more than the 6502 for example – it's just that they're not allowed to get very complicated.' Rather than waste precious silicon with a complex set of instructions, which then needed to be broken down into simple sequences that could only handle one thing at time, a RISC architecture allowed you to implement a simple pipeline and handle three different instructions simultaneously.

'You get a much simpler processor,' says Furber. 'You're much less likely to have bugs that need fixing by subsequent revisions. And so you basically got more throughput for less effort by following this RISC philosophy, and we reckoned the Berkeley people proved this pretty convincingly.'

Furber and Wilson had also convinced themselves that Acorn could design this chip – and knew that VLSI Technologies in California, which fabricated a couple of important chips in both the BBC Micro and the Acorn Electron, would be able to make it – so now they just had to convince Acorn to give them the resources.

'I didn't really believe that they could do it,' says Hermann Hauser, 'but they came to me with a design and showed me the performance figures that they were aiming for. And it was just unbelievable. It had the same number of transistors as the [Zilog] Z80 at the time, about 25,000, but 20 times the performance. It was quite extraordinary.'

Hauser famously offered Furber and Wilson 'two things that Intel, Motorola etc. didn't give their researchers,' says Wilson. 'He gave us no money and no people.'

This turned out to be a brilliant move. 'I designed fantasy instruction sets in my head,' says Wilson, 'Steve designed fantasy microarchitectures in his head. And then we'd walk down with Hermann to the pub at lunchtime, discussing them. And I'd have to give up parts of my instruction set because Steve couldn't work out how to do them. And Steve would have to make his microarchitecture better when I could convince him that we needed better execution capabilities. And that's how we made ARM.'

After a few weeks, Hauser was convinced that his two brilliant engineers could deliver on their promises. In October 1983, Acorn officially set up Project A for the Acorn RISC Machine processor (for anyone wondering, there was a Project B, but it was for the rather more mundane BBC Master).

With every single one of the 24,800 transistors in ARM1 laid out by hand, it took just over a year for the expanded Project A team – with Furber still in charge of architecture and Wilson looking after the instruction set – to create a working design. 'And all the software,' says Wilson. 'And the processor validation and verification suites. And the other three chips were also being built.'

The design was sent off to VLSI Technology in California, with sample silicon arriving back with Furber on 26 April 1985. 'The chips arrived in the morning and by the afternoon they were running BBC BASIC,' says Furber.

This quick turnaround was only possible because Acorn already had so much experience building second processors for the BBC Micro, and so that day's challenge was to slot the ARM1 chip into a pre-built prototype board and execute

Wilson's adapted version of BBC BASIC V. After the usual round of debugging, it worked, and Acorn knew it had something special on its hands: a fast yet tiny processor that also consumed a fraction of the power of similar chips.

How much power? To find out, Furber hooked up an ammeter to the board. 'I wired everything up, got the processor running code, and looked at the power,' recalls Furber. 'It was reading zero. So the processor was running code quite happily with no power.'

This wasn't, sadly, a modern-day miracle: it was a simple error on Furber's part, as he hadn't connected the power supply to the processor. It was running – quite happily – by drawing power from the surrounding, powered-up electronics (to be precise, the input/output pins had diodes to protect the chip from static electricity during handling, and from negative bus voltages during operation; these diodes were feeding power to the ARM1 chip). Still, it emphasised just how little power the ARM design needed.

In an ideal world, Acorn would have then concentrated all its ample resources on the Archimedes: the computer it had always intended to be based on the ARM. But Acorn wasn't living in an ideal world. Earlier that year, having lost millions in its bid to take the BBC Micro to the US and further millions due to all the problems with the Acorn Electron, it had almost gone into administration. Only a £12 million investment from Olivetti, in return for a 49% stake in Acorn, kept it alive.

This might have been the end for ARM, but instead Acorn took what history shows to be a brilliant, multibillion-pound decision: to ring-fence the Project A team, including the Archimedes, so that it wouldn't be cut. History suggests that its decision to continue funding the Acorn Research Centre (ARC) in Palo Alto, which had been created to develop an Archimedes operating system called ARX, was less wise.

On paper, ARX matched Acorn's ambition with ARM: not only did it offer full support for multitasking and multithreading, but it was written in Modula-2 (a programming language considered so advanced that America's Byte magazine devoted much of its August 1984 edition to it) and promised a graphical user interface to rival the Apple Macintosh.

Indeed, in the form of Jim Mitchell, Acorn could claim it had more rights to the user interface than Apple. Mitchell had worked for Xerox Palo Alto Research Center, better known as Xerox PARC, during the 1970s while the Xerox Star's famous graphical user interface had been developed. A Xerox Fellow, he was the natural person to head up Acorn's own ARC.

'It was a complex, ambitious operating system,' says Tudor Brown, the senior Acorn engineer who was essentially in charge of the Archimedes project. Or at least the hardware side of it. 'It was multitasking from the beginning and [that's] difficult to do.'

On this side of the Atlantic, development of a prototype Archimedes computer was going well. The key to this computer would be four chips, all of them developed in-house by Acorn and fabricated by VLSI Technology in Palo Alto. Brown describes one big triangle built out of four smaller triangles. At its heart, the ARM processor. On each of its sides, a video controller (VIDC), memory controller (MEMC), and input/output controller (IOC).

Early in 1986, all these chips were working and ready to slot into a circuit board designed by Brown. He soon turned this into a fully working prototype called the A500 (not to be confused with the Amiga), which each came equipped with 1MB of memory and a then-expensive 10MB hard disk. Dozens of A500 systems were shipped out to the Acorn Research Center for the developers to work on. Steve Furber and co also successfully produced ARM2, a much-enhanced version of the architecture with 27,000 transistors and even greater performance.

Meanwhile, time ticked on. When the Archimedes project was conceived, Acorn had the business market in its sights: Hauser had long wanted to build a machine that would automate office tasks, all the way back to his vision of the Acorn Proton (the design that would grow into the BBC Micro). With each passing month, however, Acorn lost further ground to IBM and the growing number of IBM-compatible PCs. While Acorn responded with its own range of Acorn Business Computers, there was no hiding from the fact that these were trumped-up BBC Micros and largely ignored by the businesses Acorn hoped to lure.

By the time 1987 rolled around, Acorn desperately needed another hit to follow up the now ageing BBC Micro. Its new masters at Olivetti were growing impatient, and so far the Project A team was only costing it money; while Acorn did release a second processor module for the BBC Micro based around ARM, it proved a hit solely with developers keen to play with this fascinating new chip.

The ARC team still hadn't delivered a working operating system, which meant Acorn now had a brilliant piece of hardware but no software to run on it. Time to call in Sophie Wilson. 'When it became obvious that the operating system written in Palo Alto wasn't going to be a success, the then director of Acorn, Brian Long, said to me, in that case, you need to write something quickly. So the operating system called Arthur, which some claim stands for Acorn written by Thursday, was concocted very rapidly. That wasn't a good operating system.'

Acorn also decided it would be wise to shift the Archimedes' focus from business to education. Time to call the BBC and see if it would lend its name to an Acorn product for the second time. Surprisingly, and controversially bearing in mind that it had no hand in the development of the Archimedes, it said yes. Even better, the BBC Archimedes soon found its way onto the list of approved computers for schools.

By June 1987, Acorn finally had a computer that it could offer for sale. And, despite all the setbacks, it wasn't being modest in its claims. A full-page ad in the national newspapers declared [1]:

Acorn are proud to announce the fastest micro in the world. The Archimedes High Performance Computer System. Using a 32-bit RISC chip, Archimedes represents a breakthrough in microprocessing technology allowing a new order of magnitude in power. Power to run programs in BBC BASIC that outperform those in assembler code on rival machines. Power to deliver 6502 and MS-DOS emulation and high level languages such as C and FORTRAN. Power to provide sound of digital stereo quality and graphics that have to be seen to believed.

To an extent, these claims were true. In its article a week after the official launch of the Archimedes, The Guardian stated [2]: 'The 2-channel, 8-voice stereo sound and graphics capabilities of the ARM are stunning.' And: 'There is also BBC BASIC V. Version IV was fast on the old 8-bit 6502 processor, and this latest incarnation is electrifying.'

But there was a big problem. Because of the late switch to a new operating system, software developers had no opportunity to create programs for this radically different platform. While it was true that Acorn had developed a DOS emulator, the company admitted to The Guardian that it was 'painfully slow'. You could use the BBC emulator to run programs developed for Micro – and note that *Elite* ran gloriously fast on the Archimedes – but by the late 1980s most Micro software was looking tired.

Then there was the small matter of price and the competition. The lowest-priced Archimedes, the 305, cost £919 once you added VAT, and only bought you 512kB of RAM. Far more sensible to buy the 310 with 1MB of RAM for £1,006. Then you probably need a monitor: Acorn's mono offering brought the price to £1,064, but a more sensible choice of colour monitor took the total to £1,236.

You would be a brave and rich parent to buy the Archimedes for your family when the Commodore Amiga and Atari ST could be bought with similar specifications for significantly less. Yes, the Archimedes was significantly faster – often, ten times or more faster – in benchmarks, but what did that matter when there was so little software to run? By this point, both the Amiga and Atari ST had built up quite a selection of games too.

Would businesses buy the Archimedes? Acorn certainly hoped so, announcing two models aimed at the business market. It claimed the 1MB-toting Archimedes 410, at a cost of £1,645 without a monitor, would ship in 1988 (this never happened), while the 440 would go on sale in November 1987 for a princely £2,704, or £2,940 with a colour monitor. In defence of this price, it had a killer specification for the time, with 4MB of RAM and a 20MB hard disk. Who knows how well it might have fared if it had launched with a suite of office programs and a killer desktop publishing offering?

By now, however, the likes of Compaq and Dell were creeping onto British shores. In the July 1987 issue of Personal Computer World, the magazine gave a glowing review to the Dell 28612 , a slick import with an 8MHz Intel 80286 processor, 1MB of RAM, a 20MB hard disk, colour monitor, 3.5-inch floppy drive, and a ready-made library of software courtesy of MS-DOS for £1,954 including VAT. 'It's not cheap in the same way that an Amstrad is cheap,' wrote Robert Schifreen [3], 'but if you want value then you've found it. For the money, it's unbeatable.'

It's telling that Acorn couldn't actually supply a full production model of the Archimedes to tie in with its official launch, but its decision to send in the highly specified A500 development system certainly gave it a chance to shine. Dick Pountain was fulsome in his praise of the machine in his preview [4]: 'The A500 felt like the fastest computer I have ever used, by a considerable margin… Just about everything you do happens instantly, and it takes a hefty-sized Mandelbrot computation before you believe that anything can tax it.'

Pountain correctly predicted that the Archimedes had its best chance of success in the education market, and if you're familiar with the Archimedes then chances are that you used it in school. Here, the BBC badge of approval and backwards compatibility with BBC Micro programs were clear plus points and, thanks to the work of Mike Muller, who would go on to co-found ARM Holdings, the Archimedes carried on its predecessor's expansion capabilities.

If you ever opened an Archimedes, you may remember a bus that ran the full width of the circuit board. This was the 'podule' bus, which allowed users to add, well, virtually anything. A second processor, Econet local networking, MIDI outputs, hard disk ports. But all at a cost, and you would have to wait for Acorn – at that point, always perilously close to bankruptcy – to build it.

Not all the promised podules would come to market, but Acorn kept on investing in the operating system. In 1989, it unveiled RISC OS 2 (a project led by William Stove), with co-operative multitasking for its windowed programs and an updated selection of core applications. What it still lacked was a wide range of third-party software. Even with the introduction of the cheaper A3000 later that year, also known as the new BBC Micro, the Archimedes never hit the highs of the first BBC Micro.

Still, the Archimedes remains one of the most notable computers of the 1980s. And it lasted much longer than some. As Hauser points out: 'It survived very successfully against the PC standard for another ten years. It was actually quite a remarkable achievement, just because it was a hit in price-performance by a mile.'

There's no doubt that many who grew up with the Archimedes feel the same affection as others towards the BBC Micro. There are just fewer of them. 'I love RISC OS and still use it today in its modern form on new ARM architecture,' said Gavin Crawford when we surveyed people about their favourite computers of the 1980s. He added: 'As a graphics artist, I used Acorn RISC PCs in my printing business. As a programmer, I still write apps for RISC OS.'

But the real impact of the Acorn Archimedes is its direct lineage towards the Raspberry Pi. Near the start of the story, we mentioned the concept of four small triangles to create one big triangle: the ARM core, video controller, I/O controller, and memory controller. Wilson believes that the ARM250, a project led by Paul Swindell and introduced in 1992, was the world's first 'system on a chip', bringing all four components onto a single die.

'I think it's quite reasonable to say that was the genesis of the idea of highly integrating things and coming up with incredibly simple single-chip computers,' says Brown. 'Fundamentally, [the Raspberry Pi] is an ARM core with integrated memory and certainly the ability to have memory directly attached. It's got I/O. It's got HDMI as the video interface. The concept of a Raspberry Pi board is as close as you can get to a single-chip computer, and that, essentially, was the genesis, the whole DNA of the Archimedes and all the ARM development.'

The Raspberry Pi was introduced in 2012. Over 30 million single-chip boards have been sold, making it the UK's most successful computer. Not a bad legacy for the Acorn Archimedes.

From Apple to Arm

Even before the Acorn Archimedes went on sale, Apple had become interested in ARM technology. VLSI, the company that fabricated the ARM chips, had been experimenting with a number of ideas, and one of these – an ARM-based graphics accelerator – had found its way to Apple HQ. 'They really loved it,' says Acorn's Tudor Brown, 'and for a while we thought Apple were going to use ARM.'

In the end, though, commercial sense took over – Apple didn't want to be dependent on Acorn, a competitor, for any of its products – and Apple switched its attention to a 32-bit microprocessor from AT&T called Hobbit. 'Long story short, in 1989 they came back to Acorn and said, look, we need this ARM thing, Hobbit doesn't work,' says Brown.

With Olivetti equally keen to divest itself of an expensive research group, Acorn, ARM, and VLSI eventually formed a joint venture called Advanced RISC Machines Ltd in November 1990. This would own all the intellectual property created by 'Project A', and its dozen employees would now work separately from Acorn. Technically, this meant neither Wilson or Furber were employees of ARM Ltd; Wilson stayed on at Acorn but was seconded to ARM Ltd 'until they didn't need me any more', while Furber left to be a university professor in Manchester (but continued to be part of the extended team as a consultant).

So why go to all this fuss? 'Apple needed a twist on the ARM, so that's why they needed to invest in it and own it,' says Brown. His new team set to work on a variation of the then-current ARM3, adding 32-bit extensions and a memory management unit (MMU). This became the ARM600, which was rapidly trimmed down in size to become the ARM610. The system-on-a-chip that would power the Apple Newton.

While the Apple Newton never became a huge commercial success, it gave a glimpse into a future of mobile devices – the Newton was powered by four AA batteries – that could act as diary, note taker, calculator… anything you can think of that could be controlled with a stylus.

Just as importantly, Apple's intervention gave ARM independence from Acorn and effectively set it free to follow its own course. This turned out to be a licensing model, so entirely different to the likes of AMD and Intel. In 1993, Texas Instruments licensed the ARM7 and within a few years the new generation of mobile phones (including the iconic Nokia 6110) had ARM processors inside.

ARM Holdings floated on the stock exchange in 1998, and never looked back. That year, it shipped 50 million chips. In 2020, the figure was well over 20 billion. If you take out the phone in your pocket, no matter who it's made by, then it's almost certain to have an ARM chip inside. 'And not just the obvious ones,' says Wilson. 'There are ARM CPUs (Cortex-M series) controlling all the subsystems, WiFi, 2G, 3G, 4G radios, storage, touchscreen, power management… You're looking at dozens of ARMs per phone.'

In 2016, Japanese telecoms company SoftBank bought ARM Holdings for $24 billion, with US chip giant Nvidia agreeing a $40 billion deal for the firm in 2020 (subject to regulatory approval). As befits such a sought-after business, in 2017 ARM rebranded itself as Arm – rather more catchy than Advanced RISC Machines Ltd.

Sources

Interviews with Tudor Brown, Christopher Curry, Steve Furber, Hermann Hauser, Andy Hopper, and Sophie Wilson.

1. Acorn Archimedes advertisement, The Observer, Sunday 21 June 1987, page 17
2. Jack Schofield and Simon Rockman, *Taking a Risc with Acorn's Educational Archie*, The Guardian, 18 June 1987, page 15
3. Robert Schifreen, Dell 28612 benchtest, Personal Computer World, July 1987, page 112
4. Acorn Archimedes A500 benchtest, Personal Computer World, August 1987, page 112

Epilogue: Whatever happened to the British PC?

The Acorn Archimedes was the last great British computer of the 20th century. Yes, there were many fine Windows-based computers – even today I recommend, in my day job as editor of *PC Pro* magazine, that people's first port of call for a desktop computer is one of this country's fine bespoke PC manufacturers. But they are using off-the-shelf cases made in China filled with components sourced from far-flung lands. No matter how hard you try, there's no hiding from the fact that the mainstream is now dominated by American, Chinese, and Taiwanese companies.

To fully understand why would take a whole new book, but the two key factors behind this shift are scale and standardisation. As soon as IBM entered the personal computer market in 1981, the writing was on the phosphate-coated screen. To compete, you either needed to produce such an amazingly compelling platform that people would rush to write software and games for it, or you needed to jump on the same platform as IBM and somehow differentiate your computer from all the others.

Let's take that first possibility: creating your own platform. It turns out this is quite tricky. Even Apple struggled to compete against Microsoft until Steve Jobs returned from exile in 1996. You can argue that Google's success with Chromebooks shows there is room for a third player, but this gives us a clue: Google is a global power so far beyond the scale of Acorn, Amstrad, and Sinclair that this is probably the first time that all four have been included in the same sentence.

It's telling that only Amstrad, with its entrepreneurial owner, managed to make any impression in the States. Attempting to bring the BBC Micro to the USA almost bankrupted Acorn, while Sinclair took the more cautious approach of partnering with Timex in its cross-Atlantic adventure. In truth, none of the British companies were more than bit players in a market full of giants. Even Commodore, a billion-dollar business by 1984, was soon crushed by the weight of IBM clones.

Among this IBM PC-compatible gloom there stood just one British rebel: Acorn. Despite being owned by Olivetti at the time, it persuaded the BBC to take a gamble on the Acorn Archimedes with its revolutionary ARM processor. The Archimedes battled on for over a decade before eventually bowing to the inevitable.

This is why the Brits all tried their hands at creating IBM-compatible PCs in the late 1980s – including Amstrad. But how do you differentiate yourself when each computer is essentially the same? Again, British companies could not match the scale of the invading American companies: Dell's aggressive pricing and 'just in time' delivery model proved decisive in many British companies' downfall.

There were success stories, and these were mostly built around education. Viglen – a company Amstrad bought in 1994 before selling it two decades later – and Research Machines continued to build their own computers until a few years ago, with their 'value add' being the support and advice they gave to schools. They provided solutions to problems, not boxes with components inside. And still do.

Then, in 2012, amidst this barren British wasteland, a green shoot. Once more, it emerged in Cambridge, this time the brainchild of research student Eben Christopher Upton. With plenty of encouragement from Jack Lang – a serial entrepreneur himself, but crucially a long-time lecturer at the University of Cambridge's Computer Laboratory – he co-founded the Raspberry Pi Foundation to create low-cost, single-board computers.

The first Raspberry Pi went on sale for £22 in February 2012, and within eight years it has become the best-selling British computer of all time: on 14 December 2019, Upton casually tweeted that 'we sold our thirty-millionth unit some time last week (we think Tuesday)'.

That means that Raspberry Pi is now officially the best-selling computer of all time. So, whatever happened to the British PC? It's here, and more successful than ever. Let's hope that it inspires another golden generation of British coders, which is exactly what the Raspberry Pi Foundation was set up to do.

Acknowledgements

To say this book is a group effort is an understatement roughly equivalent to 'the BBC Micro was quite an important British computer'. If you'll forgive the computing metaphor, the output you see before was only made possible through the input of hundreds of people.

First, I must thank the team at Raspberry Pi Press who made this all happen. Top of the list: Simon Brew. He was the one who casually asked if I had any ideas for a book, which at that point was only a title. He then enthusiastically sold the idea to the head of Raspberry Pi Press, Russell Barnes. Together with Simon and Russell we took that idea and turned it into an outline that could actually become a book.

My thanks also to all at the National Museum of Computing, but particularly to Jacqui Garrard for her boundless energy and bearing with me as I sat in the café and library poring through back issues of Personal Computer World (or PCW as it was affectionately known).

Then came the work of identifying which computers to cover. Sales figures of computers in 1980s Britain are difficult to pin down. With the help of hundreds of people on Twitter, we squeezed a long list of around 100 to the 19 in this book. So thank you to everyone who took the time to fill in our initial survey.

With the short list identified, I was keen to speak to as many people who were involved as possible. My 'in' to this rarefied world was Jack Lang, a co-founder of the Raspberry Pi Foundation, serial entrepreneur, and long-time lecturer at the University of Cambridge's Computer Lab. He put me in touch directly with a number of people you see quoted in these pages, and that was enough to set the ball rolling. It helps that one person knows another, who knows another, who – you get the idea. I would also like to thank Peter Rennison, who put me in contact with people even Jack couldn't reach.

I can't thank my interviewees enough. Without exception, they willingly gave up their time to relive the birth of the computers they created. With some, it was a 20-minute conversation. At the other extreme came wide-ranging discussions that lasted four hours, split over a couple of days.

And what a list of interviewees: Tudor Brown, John Caswell, Charles Cotton, Christopher Curry, Mike Fischer, Steve Furber, Richard Harding, Neil Harris, Hermann Hauser, Andrew Herbert, Andy Hopper, Jan Jones, Mark-Eric Jones, David Karlin, Tim King, John Mathieson, Richard Miller, Mike O'Regan, Roland Perry, Nigel Searle, Kit Spencer, Ian Thompson-Bell, Michael Tomczyk, Leonard Tramiel, Chris Turner, and (last, but not by any measurable means least) Sophie Wilson.

Thank you not only for your time in the interviews, but also for being kind enough to check through the final draft of this book.

I also interviewed two journalists whose names kept on popping up when I read through contemporary newspaper reports and magazine write-ups. Dick Pountain, a legend from his days at Byte Magazine and then PCW, who would go on to co-found PC Pro magazine. Thank you, Dick, for your ridiculous patience on the phone as we went through the early years of your extensive career.

I also spoke at length to the long-time Computing editor of The Guardian, Jack Schofield, who passed away in March 2020. Over the past few months, I've read countless of his articles from the 1980s, and his knowledge and sheer wisdom are weaved into each one. He continued to provide sage advice to his legions of readers in his weekly *Ask Jack* column. To say he will be much missed doesn't begin to cover it.

Special thanks must go to Rupert Goodwins. Rupert is that rarest of creatures: not only is he incredibly technical and a great writer, he's also extraordinarily – and I hate to use this word – nice. I gave Rupert an early copy of this book for his interest and to check the areas where I quoted him. He then spent his own time going through it meticulously, spotting errors and calling me to account in a way that I will be eternally grateful for. While I am entirely responsible for any mistakes that have slipped through, Rupert caught dozens of assumptions and inaccuracies that I could never have spotted. Thank you, Rupert.

Finally, I would like to thank my family. Kim, Fraser, Rowan, and Penelope all put up with me working late into the night for months, as I attempted to haul the myriad sources into something vaguely resembling a book. Thank you for your patience and good humour throughout. I also need to thank my father, Stuart Danton, without whom I wouldn't be writing these words at all. After all, it was him who photographed me in front of an MSI 6800 System One micro at the age of six, playing Roger Conway's Game of Life, and resulted in my first appearance in a computer magazine – as part of his review of the MSI computer back in the June 1979 issue of Personal Computer World.

Further reading, further viewing and forums

There are numerous excellent online resources and books available if this has made you curious to read yet more. While many of the books here are no longer in print, several are still available as e-books, PDF downloads, and on archive websites.

General background

- Christopher Evans, *The Mighty Micro: The Impact of the Computer Revolution*, 1979, Victor Gollancz, ISBN 0575027088

- Paul Freiberger and Michael Swaine, *Fire in the Valley: The Birth and Death of the Personal Computer*, 1984, McGraw-Hill, ISBN 1937785769

- Tom Lean, *Electronic Dreams: How 1980s Britain Learned to Love the Computer*, 2016, Bloomsbury Sega, ISBN 9781472918338

- Gordon Laing, *Digital Retro: The Evolution and Design of the Personal Computer*, 2004, Sybex, ISBN 078214330X

Acorn/BBC Micro

- Tilly Blyth, *The Legacy of BBC Micro*, Nesta 2013, **nesta.org.uk/report/the-legacy-of-bbc-micro** (free PDF download)

- BBC Computer Literacy Project 1980-1999, **clp.bbcrewind.co.uk**

Amstrad

- Alan Sugar, *What You See Is What You Get: My Autobiography*, Pan Books 2011, ISBN-13 978-0330520478

- David Thomas, *Alan Sugar: The Amstrad Story*, Century 1990, ISBN 0712635181
Amstrad CPC 464 group on Facebook

Apple

- Brent Schlender and Rick Tetzeli, *Becoming Steve Jobs: The Evolution of a Reckless Upstart into a Visionary Leader*, 2015, Crown Business, ISBN 0385347405

- Walter Isaacson, *Steve Jobs*, 2011, Simon & Schuster, ISBN 1451648537

- Andy Hertzfeld and others, Folklore.org, "Anecdotes about the development of Apple's original Macintosh, and the people who made it"

Commodore

- Brian Bagnall, *On the Edge: The Spectacular Rise and Fall of Commodore*, 2005, Variant Press, ISBN 0973864907

- Brian Bagnall, *Commodore: The Amiga Years*, 2012, Variant Press, ISBN 0973864990

IBM

- James Chposky and Ted Leonsis, *Blue Magic: The People, the Power and the Politics behind the IBM Personal Computer*, ISBN 0246134453

Sinclair

- Rodney Dale, *The Sinclair Story*, 1985, Duckworth, ISBN 0715619012

- Ian Adamson and Richard Kennedy, *Sinclair and the Sunrise Technology*, 1986, Penguin, ISBN 0140087745

Archive magazines

If you would like to retrace the early history of computers, and have a lot of time on your hands, then there's no match for reading through archive issues of magazines. You can read many online via the Internet Archive, but for the 'real thing' head to The National Museum of Computing, situated on the same site as Bletchley Park just outside Milton Keynes.

Archive newspapers

This book could not have been created without access to online newspaper archives. Many can be found via your local library – simply log on to its online service and check what's available – but others (such as the Financial Times) are largely restricted to academic institutions. This book also relied upon the international archive available at **newspapers.com** (a paid-for service).

YouTube

If you want to see British heroes of computing being interviewed in the flesh, The Centre for Computing History has a YouTube channel that features everyone from Steve Furber (of Acorn fame) to Rick Dickinson (the late designer of Sinclair's most successful computers).

For an American point of view, head to a similar series of videos by the Computer History Museum. Perhaps most valuable are its series of oral histories, including those of Chuck Peddle (chief architect of the Commodore PET) and Pat Gelsinger (current CEO of Intel).

Films

The Commodore Story, 2018, directed by Steven Fletcher

Micro Men, 2009, directed by Saul Metzstein

Steve Jobs: The Man in the Machine, 2015, directed by Alex Gibney

Index